W9-CUO-583

The
NexStar
User's Guide

Patrick Moore's
Practical
Astronomy
Series

CELESTRON

The NexStar User's Guide

Michael W. Swanson

With 115 Figures
(including 22 in color)

Springer

NexStar is a registered trademark of Celestron. TheSky is a registered trademark of Software Bisque. All other proprietary names are the property of their respective companies.

British Library Cataloguing in Publication Data
  Swanson, Michael W.
  The NexStar user's guide. – (Patrick Moore's practical astronomy series)
  1. Telescopes
  I. Title
  522.2
ISBN 1852337141

Library of Congress Cataloging-in-Publication Data
Swanson, Michael W., 1963-
  The NexStar user guide/Michael W. Swanson.
    p. cm. – (Patrick Moore's practical astronomy series)
  Includes index.
  ISBN 1-85233-714-1 (alk. paper)
  1. Telescopes. I. Title. II. Series.
QB88.S88 2003
522'.2–dc21   2003050549

Apart from any fair dealing for the purposes of research or private study, or criticism or review, as permitted under the Copyright, Designs and Patents Act 1988, this publication may only be reproduced, stored or transmitted, in any form or by any means, with the prior permission in writing of the publishers, or in the case of reprographic reproduction in accordance with the terms of licences issued by the Copyright Licensing Agency. Enquiries concerning reproduction outside those terms should be sent to the publishers.

Patrick Moore's Practical Astronomy Series ISSN 1617-7185
ISBN 1-85233-714-1 Springer-Verlag London Berlin Heidelberg
Springer Science+Business Media
springeronline.com

© Springer-Verlag London Limited 2004
Printed in Singapore
2nd printing 2005

The use of registered names, trademarks, etc. in this publication does not imply, even in the absence of a specific statement, that such names are exempt from the relevant laws and regulations and therefore free for general use.

The publisher makes no representation, express or implied, with regard to the accuracy of the information contained in this book and cannot accept any legal responsibility or liability for any errors or omissions that may be made. Observing the Sun, along with a few other aspects of astronomy, can be dangerous. Neither the publisher nor the author accept any legal responsibility or liability for personal loss or injury caused, or alleged to have been caused, by any information or recommendation contained in this book.

Typeset by EXPO Holdings, Malaysia
58/3830-54321   Printed on acid-free paper   SPIN 11331179

Contents

Contents

List of Tables

List of Figures

Acknowledgements

It is true with all collection of information that the collector didn't collect it in a vacuum. Much of the knowledge in this book came into being via discussions with other owners of NexStar telescopes, most of whom deserve more than the simple acknoledgement I offer here.

Thanks to all members of the NexStar and NexStarGPS discussion groups on Yahoo, especially Bob Berta, Kevin Brett, John Hilliard, Rod Mollise, Jeff Richards, Joe Shuster, and Hank Williams. Refer to the internet resources in Appendix A to join us in our discussions.

I would like to thank John Carlyle, Alain Fraysse, Carroll Morgan, and Jean Piquette for discussions and their work on standardizing and improving the alignment procedure for various models of NexStar telescopes. The fruits of my association with these gentlemen can be seen in Chapter 4.

Special thanks to Matthias Bopp, Ray Cooper, Phil Chambers, Dan Hupp, and Mario Mariani for their input on mechanical, optical, and other issues. Thanks once more to Ray Cooper and also Tamás Székffy whose support in providing server space for file downloads has made the NexStar Resouce Site a success.

I have also been fortunate to have direct access to several wonderful people at Celestron who provided many technical details necessary to make this book a more valuable resource. Thank you to the management at Celestron for making this possible.

Extra special thanks go to Frank Dilatush for proofreading the book, offering suggestions, and providing great support to all members on the NexStar Yahoo Group. When I was close to pulling my hair out, Frank provided assistance, encouragement, and frequent comic relief.

I also thank my wife, Yumiko, for supporting and encouraging me in all that I do and putting up with the tap, tap, tap of keys at all times day and night. When I was frustrated or lazy, it was often her determination that kept me going. When I was too tired to think, it was her common sense that convinced me to put the work aside. I'm not sure what I did to deserve such a wonderful companion, but I am thankful everyday that I am so fortunate.

This book is dedicated to my grandfather, Jim Swanson, and my father, Chuck Swanson, for my first look through a telescope when I was five years old — a glorious view of our Moon while Dad helped me up to the eyepiece of Grandpa's white-tube refractor. That moment sparked a lifelong curiosity and affection of the sky for which I am eternally grateful.

Chapter 1

Introduction

It was a dark and stormy night ...

Thanks to Charles Schultz (and Snoopy), it was guaranteed that if I ever wrote a book, that would be the opening line. And as things would have it, that opening also describes the first night after I received my new NexStar telescope. But naturally that couldn't last for long and my little NexStar 80 GT soon saw first light. After an interlude of about four years from my favorite hobby, I was testing the waters with one of the new technological wonders from Celestron.

Would reality match the hype? Would the GoTo and tracking features work as expected? I was soon to find out.

Assembly was simple. The alignment procedure seemed straightforward. After just a few moments I was ready for the test. Press the Planet button, scroll to Jupiter, press the Enter button and away it went. As it slowed to center the target, I put my eye to the eyepiece. Stars drifted slowly by until, BANG, there it was: Jupiter! Not precisely centered, but certainly close enough for my first try.

But not all nights were so successful. Sometimes the target would not be in the field of view. I deliberately set out to improve my results and get the most out of all this technology. This book is the outcome of that deliberate and sometimes frustrating journey.

I enjoyed that first NexStar so much that I also now own a NexStar 11 GPS and a NexStar 114 mount that is outfitted to carry a variety of optical tubes.

And how does that N80GT behave these days? Every object in the eyepiece, every time.

Are GoTo Scopes Appropriate for Beginners?

Many seasoned amateur astronomers have shown disdain with the advent of entry-level computerized GoTo telescopes, even though many of them own more expensive, advanced scopes of the GoTo variety. They make the criticism that one of the joys of astronomy is learning our way around the sky and that the money spent for the computerization would be more wisely applied towards better optics. These seem to be valid points, but the more experience I have with budding amateur astronomers and these new scopes, the more I tend to disagree.

First, it is a mistaken assumption that using a computerized scope alleviates the need to learn the night sky. It is still necessary to look at star charts or fire up a planetarium program in order to decide what to look at. The beginner still learns the sky; they just don't learn to star hop – jump from one known star to another until they have found the quarry. Granted, some will not follow through in acquiring even this computer-assisted knowledge, but would they have taken the much steeper path learning to star hop? Especially now that most of us live under urban light domes with so few guide stars? I see beginners with GoTo scopes having a higher rate of continued interest than beginners with more conventional introductory models. Could it be that the GoTo beginners are seeing more and it keeps them interested?

In most cases, these entry-level computerized scopes cost about $100–150 more than a non-motorized scope of similar quality. Throw in the usual $60 clock drive for a non-computerized scope and the price difference doesn't amount to much. I will agree that the least expensive of these scopes are suspect as astronomical instruments, but in particular I find most of the 80mm and larger scopes quite up to the task of starting an amateur astronomer on their lifetime journey.

This new breed of entry-level telescope has gained popularity unlike anything in amateur astronomy that came before. With this comes an unprecedented, widespread awareness about the wonders of astronomy. In the end, astronomy is about expanding our awareness by learning and observing. These are phenomena a computer-assisted tourist can experience just as well as the star-hopping navigator. We need to embrace this expansion of technology and guide beginning amateurs in appropriate new ways. I hope this book is a step in that direction.

A Brief History of NexStar

A newcomer to amateur astronomy would think that inexpensive, entry-level GoTo scopes have been the norm for years. The advertisements from Celestron and Meade, as well as all of their resellers, make it seem so. It is certain from all the discussions in the readers' forums and on the Internet that it is one of the most important areas of development since the introduction of quality, mass-produced scopes in the 1960s. But actually, inexpensive GoTo scopes are a new development. And just what is a GoTo scope?

GoTo scopes have several common characteristics. First, they include motor drives on both axes to allow the scope to be pointed at any desired location. Second, they include encoders to keep track of the location of each axis. Third, they have a handheld computer (computerized hand control) that contains a database (catalog) of astronomical objects and performs the calculations necessary for locating these objects. And finally, the computer control instructs the motors to move the scope to the desired object and automatically tracks its motion through the sky.

The hand control is able to find objects in the sky after it points to two bright "alignment" stars, which you then center in the eyepiece. NexStar telescopes can either locate those two alignment stars for you or you can locate them yourself. To locate them for you, the scope must know its location on the Earth, the current date and time, and the direction of North. The NexStar GPS models get this information from the Global Positioning System satellite network and an internal compass; for all other models you must provide the information each time you perform the alignment.

GoTo features have been available to professional astronomers for years, although only on observatory telescopes and not in a portable fashion. More recently, beginning in the early 1990s, many amateur astronomers could have such features in the form of the

Figure 1.1. An early NexStar 5 advertisement.

Celestron Ultima 2000 and the Meade LX200. But the $2000 to $3000 dollar price (in the US) could not be considered "GoTo for the masses".

Then came the late 1990s and Meade's introduction of the ETX-90EC. For about $750 (US), you could now own a full-fledged GoTo scope with 90mm Maksutov–Cassegrain optics. Meade had made attempts at entry-level GoTo scopes before, but this one was the

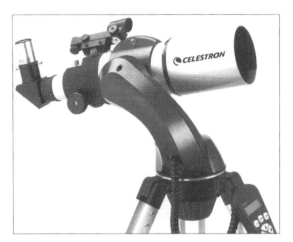

Figure 1.2. My start in GoTo scopes – the NexStar 80 GT. Photo courtesy Celestron.

first true success. By the end of 1999, Meade also announced the ETX-125EC, 5 inches of aperture for an additional $300. *Sky and Telescope* made a prediction in May of 1999: "The ETX/Autostar concept will go down as the greatest happening in amateur astronomy yet."

Celestron answered with the NexStar 5. Announced in the summer of 1999, its physical appearance was unique to say the least. A single sculpted arm, seeming more at home in a museum of modern art than on top of a tripod, was the only support for the long-admired Celestron 5-inch Schmidt–Cassegrain optical tube. Immediately the question arose whether a single fork arm was sufficient to support the tube. Owners of the soon-to-be-discontinued C5+ already knew the answer: yes. In fact, comparisons would soon confirm that the NexStar 5 was more stable than the ETX-125EC. The NexStar 5, including computerized hand control, was released at the price of $1199, so it needed something to recommend it above the ETX. In addition to the added stability, the N5 also featured all-metal construction in all the right places. There was no plastic in the supports and drive train as was found in the ETX series.

Another question was raised as soon as the first N5s found their way into owners' hands: why were the fork arm and drive base so massive for just a 5-inch optical tube? It was obvious that Celestron had something else in store. We found in early 2000 just what that was: an 8-inch version called – well, what else? – NexStar 8. Reaching owners in the late fall, the single fork arm was still a great support for the added weight. And the introductory price of $1899, including tripod and computerized hand control, offered excellent value for the features.

Nonetheless, these scopes were not truly GoTo for the masses. The announcement for the rest of us came in the summer of 2000. Meade introduced two little ETX refractors – 60mm and 70mm models. At just $299 and $349, including computerized hand control (but still lacking a tripod), now almost anyone could afford a GoTo scope. Celestron answered a month later with the NexStar 60 and 80 (refractors) and the NexStar 114 (Newtonian reflector). Versions with and without computer hand control were available, but almost all units were sold with the computer – the GT models. The GT models were introduced at prices from $300 to $500 and all included a tripod.

By the end of 2000, Celestron introduced the NexStar 4 GT, a 4-inch Maksutov–Cassegrain on a GoTo mount. Selling for about $700 at release, it was a strong competitor in the small scope arena and an instant success. Although the NexStar 4 looks like its cousin the NexStar 5, actually it is more closely related to the GT models. They share similar motors and gears (less powerful than the NexStar 5 and 8) and the same hand control. This scope was truly designed to compete with the Meade ETX-90EC. The NexStar 4 has more aperture, is mechanically much sturdier than the ETX-90EC, and costs less after adding the computerized hand control to the ETX.

Perhaps the most exciting development of the NexStar line came with the announcement of the NexStar 11 GPS in the spring of 2001. Unlike all previous NexStar scopes, the N11GPS was designed with astrophotography in mind. Optically, the N11GPS is the same as the renowned C11 with the distinction of a carbon fiber tube, rather than aluminum. As the temperature changes, carbon fiber expands and contracts much less than aluminum, a feature critical to long-exposure astrophotography. During a long exposure, an aluminum tube can contract enough as the night becomes cooler to change the focus of the scope and thus ruin the image. Additionally, the N11GPS sports the Fastar optic system to allow digital imaging at a super-fast focal ratio of f/2. The drive base and fork mount are rock-solid and provide smooth tracking.

But the feature that captured the imagination of many was integration of Global Positioning System (GPS) technology. Using a GPS receiver and a mechanical leveling sensor, the NexStar 11 GPS made the initial alignment process even easier. You simply set up the tripod, bolt the scope on top, power up, and choose GPS alignment. The scope points roughly north and level, links to GPS and slews roughly to the first alignment star. You center the alignment star, push a button and off it goes to the second alignment star. You center the second star and you are ready to go. Elapsed time can be as little as 2 minutes. The release of the NexStar 11 GPS was one of the most anticipated shipping dates for a telescope in recent memory.

Hot on the heels of the N11, Celestron released the NexStar 8 GPS. Sporting all of the optical, mechanical, and electronic features of its big brother, the N8GPS is a more portable telescope for those not needing the larger aperture of the N11.

And finally, in the spring of 2002, Celestron announced their new NexStar models, the 5i and 8i. With the same basic mechanical design as the earlier NexStar 5 and 8, the N5i and 8i allow you to start with a basic telescope and add components as you desire. They come standard with a non-computerized, electronic hand control that allows pushbutton directional control. By adding the computerized hand control you get GoTo, tracking, and most of the features of the NexStar 8 and 11 GPS models. Add the CN-16 GPS module and you get a GPS receiver, auto-leveling, and an electronic compass that allows GPS setup as found on the N8/11GPS. The NexStar 5i and 8i entered the market in the summer of 2002 and are proving to be great additions to the NexStar line.

The NexStar Resource Site

The origins of this book trace back to the NexStar Resource Site – http://www.NexStarSite.com. I began this site soon after I began using my NexStar 80 GT (Figure 1.2) and noted there was a wealth of information being discussed on the Yahoo NexStar discussion group. It was obvious that a web site would be the perfect means to collect and publish the hard-learned lessons of the NexStar Group. Many of the tips found throughout this book and in Chapter 14 in particular are found in the "Odds and Ends" section of the site . New tips are added regularly, so if you have a question or problem not addressed here, perhaps the solution will be found on the NexStar Resource Site.

Drawing on my background in computer programming and database design, I soon started adding various resources to the site, all free for the taking. One particular download is by far the most popular: NexStar Observer List (NSOL). NSOL is session planning software with the capability of controlling any of the NexStar models via the serial port of a Windows-based computer. Thousands of NexStar owners have downloaded NSOL and Celestron now provides it with all new NexStar 60, 80, 114, and 4 telescopes. NSOL is discussed in detail in Chapter 7 and the latest version is available free for download from the NexStar Resource Site.

In fact, several of the resources discussed in this book are available for download at the site (Figure 1.3). With the publishing of this book I have created a separate section that lists these resources organized by chapters corresponding to the book. Direct links to all of the Internet resources in Appendix A are also included. And hard as I have tried, undoubtedly a few errors have found their way into print in these pages. As they are discovered, corrections to these errors will be published on the web site as well.

Please feel free to visit the web site; and I always answer all email I receive at swanson.michael@usa.net.

Organization of the Book

Chapter 2 – "Astronomy Basics" – is an introduction to the world of amateur astronomy. Concepts presented here will provide the background knowledge required to become more than just a "button pusher".

Chapter 3 – "Overview of the NexStar Line" – provides information on each NexStar model to assist you in choosing the NexStar that best suits your needs.

Chapter 4 – "Alignment" – details the various star alignment procedures insuring your scope delivers accurate GoTo performance. If you are having problems getting your NexStar to point accurately, this chapter will have it whipped into shape in no time.

Chapter 5 – "Basic Operation" – is a detailed presentation of the operation of a NexStar telescope. If the manual provided with the telescope leaves you wondering, this is the place to start.

Chapter 6 – "Expanding Your Horizons – Choosing Objects to View" – is written for the new amateur astronomer. After the Moon, planets, and built-in Tour, how do you decide what to view?

Chapter 7 – "Using the Software Included with the Little NexStars – TheSky and NexStar Observer List" – is your guide to using the PC software that comes with the little NexStars. Various other models of NexStars have included TheSky on occasion as well.

Chapter 8 – "Accessories for Your NexStar" – provides recommendations on tripods, eyepieces, power sources, and more. Several accessories and modifications are projects you can do at home for little cost.

Chapter 9 – "Collimation – Optical Alignment" – provides the details you need to properly align your telescope's optics to insure the best possible optical performance.

Chapter 10 – "Controlling Your NexStar with a PC or Palmtop Computer" – details the sometimes confusing requirements for interfacing NexStar telescopes with a computer to allow advanced control via specialized astronomy software.

Chapter 11 – "Astrophotography with a NexStar" – is an introduction to astrophotography with NexStar telescopes, from the simplest methods that most anyone can afford and find successful, to an overview of the world of the serious astrophotographer.

Chapter 12 – "Maintenance, Care, and Cleaning" – shows how, with a little care, you can keep your NexStar running like new for years to come.

Chapter 13 – "Mounting Other Optical Tubes on a NexStar" – gives details on methods and accessories to attach different optical tubes to most NexStar mounts.

NexStar Resource Site

Home | NexStar 50 Club | Tracking Test | Downloads | PC Control | Useful Links
NexStar Alignment Guide | Equipment Reviews | Odds and Ends

Tip: Do **not** close the ad window and it will remain in the background while you visit other pages!

Welcome to the NexStar Resource Site!
Home of NexStar Observer List
and THE Internet Resource for All Models
of NexStar Telescopes

Email this site to a friend!

NexStar Observer List is a **free** PC control program for all models of Celestron NexStar and Tasco StarGuide scopes. Version 2 was released December 17, 2001, read more and download your copy in the Downloads section.

The NexStar line of telescopes is the newest offering from Celestron International. Most of the models feature a computer hand controller that allows the scope to automatically "go to" objects in its generous database. They track objects automatically as well. If you haven't heard much about the NexStar line, I recommend you visit the Celestron web site in "Useful Links" above. Then you might find it useful to listen to the discussions at Yahoo Groups, also on "Useful Links". The folks on the NexStar Group are quite friendly and don't mind questions from someone trying to make up their mind about a purchase and will help you get the most out of your NexStar telescope.

Clear Skies!
Mike Swanson
swanson.michael@usa.net

What's New?

- Odds & Ends updated 24 Dec - Solutions for problems with GPS links
- Odds & Ends updated 17 Dec - Link to Matthias Bopp's internal hand control heater
- PC Control updated 16 Dec - STAR Atlas PRO and Voyager III with SkyPilot both support NexStar telescopes
- PC Control updated 3 Dec - Added "Arrow Keys for GuideStar" a unique package allowing remote control of the little NexStars
- Tracking Test updated 26 Nov - Added info about GPS and N5i/8i tracking
- Odds & Ends updated 14 Nov - Info regarding possible problems encountered after motor control board replacement on N11
- Downloads and PC Control updated 14 Nov - Added a link to a list of the most common problems encountered when controlling a scope with a PC

Copyright 2002,
Michael Swanson

Contact the webmaster
swanson.michael@usa.net

Figure 1.3. The NexStar Resource Site – http://www.NexStarSite.com.

Chapter 14 – "Additional Tips and Solutions" – is a collection of specific tips not covered elsewhere in the book.

Appendix A – "Internet Resources" – is a short guide to the wealth of information available on the Internet.

Appendix B – "Objects in the NexStar Hand Control" – provides a rundown on the objects in the computerized hand control of the various models to include a complete compilation of the Named Object, Asterism, and other unique NexStar lists.

Appendix C – "PC and Palmtop Software Compatible with NexStar Telescopes" – lists programs compatible with NexStar telescopes at the time of this writing.

Appendix D – "Writing Programs to Control NexStar Telescopes" – is a specialized section for those interested in writing software to control a NexStar telescope with personal computers and palmtop devices.

Appendix E – "Glossary" – should include any unfamiliar term you may run across as you are reading.

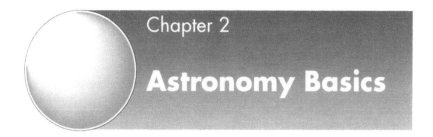

Chapter 2

Astronomy Basics

Before we continue it is important to have a basic knowledge of general astronomy. This chapter provides just that type of information; we will pick up again with NexStar telescopes in the next chapter. Let's avoid the heavy theory and complex diagrams – think of this as "Astronomy Light" (no pun intended).

Astronomy is arguably humankind's oldest science. Historic and prehistoric accounts show that as long as humans have been recording things important to them, the sky has figured prominently. While we have been able to very accurately determine the motion of the objects in the sky for more than three thousand years, it is only a relatively recent development that we have come to understand why they move the way they do. Unfortunately, most folks still don't know the why and don't even notice the motion!

The Night Sky

After you spend a little time under the night sky, you begin to notice things that were not immediately apparent. As the night passes, the various star patterns drift slowly overhead, coming up from the east and setting in the west. The stars differ in brightness and seem to form recognizable patterns. Some nights are clearer than others, even comparing various nights with no clouds. Occasionally a bright light passes overhead unexpectedly. Night after night, the phase and location of the Moon change dramatically. There is a lot to be seen if you are observant. At first it can be quite confusing, but there are some simple concepts that can help as you slowly start to make sense of it all.

Constellations

Stars are so far from us that their motion from year to year is almost negligible. The patterns that you come to recognize will remain virtually unchanged for hundreds of years. The planets and other solar system objects wander around the sky, but the stars stay relatively fixed in relation to one another.

Many star patterns have names and are known to even the most casual observer of the night sky. For instance, in the spring, observers in the Northern Hemisphere easily identify the group of seven stars known as the Big Dipper. During the months from October through February, the hourglass shape of Orion is readily visible to observers in both the Northern and Southern Hemispheres. The Big Dipper constitutes the brightest stars in the constellation Ursa Major – the Big Bear. Orion, the Hunter, is a constellation in its own right. There are 88 constellations that professional astronomers established to separate the sky into regions in the same way that the Earth is separated into continents and oceans.

The Motion of the Sky

The Big Dipper holds another distinction besides being the most recognizable constellation in northern skies. The two stars at the end of its bowl point directly to Polaris, the North Star. Located almost directly over the Earth's North Pole, Polaris is the pivot point that the sky seems to swing around as the night goes on. Unlike all other objects you see in the sky, Polaris stays put! And as its name implies, it unfailingly points the way north. When you face north, all the stars around Polaris travel in a counterclockwise circle. The point that Polaris marks is known as the north celestial pole. Observers in the Southern Hemisphere see a mirror of this when they look to the south. Stars travel in a clockwise circle around the south celestial pole, although there is no bright star to mark that point.

The situation is different when we look away from the poles. When northern observers look south, or when southern observers look north, they will notice that the stars rise in the east and set in the west, just as the Sun does each day. All of this continuous motion is caused by the Earth's 24-hour daily rotation. The Earth spins on its axis of rotation once every 24 hours, causing the day and night as well as the moment-by-moment drift of the objects in the sky.

In addition to rotating on its axis, the Earth also makes a long elliptical journey around the Sun. This trip takes about 365 days – one year. The stars and constellations are basically frozen in their relative locations but the Earth's movement around the Sun causes the constellations to drift a bit further to the west, night after night. The stars we see at night are those on the side of the Earth away from the Sun, as shown in Figure 2.1. The constellations are always in the same place, but as the year progresses, the stars we were viewing a

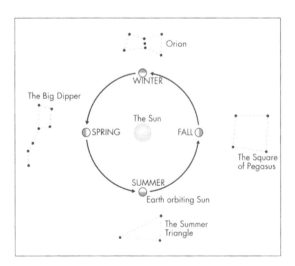

Figure 2.1. As the Earth travels around the Sun throughout the year, different sections of the sky are visible at night. This figure depicts the major star formations visible in the evening; the seasons indicated are in reference to the Northern Hemisphere.

few months ago are in our daytime sky. For example, in the winter, Orion is prominent in our night sky. But in the summer, Orion is behind the Sun, or in other words, it is in our sky during the day when the brilliance of the Sun hides the stars from our view.

Sky Coordinates

Besides the constellations, we also refer to other imaginary boundaries in the sky. The horizon is the line where the land meets the sky. The zenith is the point directly overhead. The meridian is the line running from the northern horizon, up through the north celestial pole, overhead through the zenith, then down to the southern horizon. Thus it splits the sky into eastern and western halves. The celestial equator is a line that runs from east to west, directly above the Earth's equator. And finally, the ecliptic is a wavy line traveling north, then south of the celestial equator. The ecliptic is significant as the Sun, the Moon, and all the planets travel through our sky near to this line.

Just as longitude and latitude are used to pinpoint locations on the Earth, we use right ascension (RA) and declination (Dec) to pinpoint locations in the sky. As shown in Figure 2.2, lines of right ascension run from the north celestial pole to the south celestial pole, like longitude on the Earth. Thus they meet or converge at the celestial poles. Lines of declination run east to west, parallel to one another, just like latitude.

We measure right ascension in hours, minutes, and seconds. RA starts at $0^h\,00^m\,00^s$ then goes clockwise around the north celestial pole until we come to $23^h\,59^m\,59^s$ just before

Figure 2.2. Lines of right ascension around the north celestial pole. Created in Patrick Chevalley's Cartes du Ciel (Sky Charts).

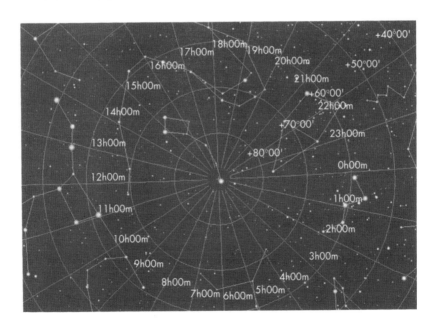

where we started. Thus there are 24 hours of right ascension. Declination is measured in degrees (°), arc minutes (′), and arc seconds (″). The declination of the celestial equator (right above the Earth's equator) is 0° 00′ 00″ (0°), the declination of the north celestial pole is 90° and the declination of the south celestial pole is –90°. From this system, we can give the coordinates for any object in the sky. For example, the coordinates for Rigel, a bright star in the constellation Orion, are RA 05^h 14^m 30^s , Dec –08° 12′ 06″.

The line of right ascension directly above us at the meridian is known as local sidereal time (LST). Every hour, local sidereal time changes about one hour. In other words, if local sidereal time is currently 18^h RA, in one hour LST will be 19^h RA. Naturally this corresponds to the fact that the Earth rotates once every 24 hours. Sidereal rate is the rate at which objects move across the sky – approximately one hour of right ascension for every hour of time here on Earth. Since the 360° of the circle divided by 24 hours yields 15, this rate of motion corresponds to 15° at the celestial equator.

Measuring "Distance" Between Two Objects

We measure the "distance" between two objects as the angular separation between them. If you project a straight line from you to each object and measure the angle between the two lines, that is the angular separation. We express this as degrees, arc minutes, and arc seconds. The angular separation between Dubhe, the bright star at the end of the Big Dipper's bowl, and Polaris, the North Star, is about 28° 42′ 30″ or nearly 30°.

Estimating angular separation when you are outdoors is quite easy. Hold your hand up at arm's length and your little finger covers about 1 degree. Your index, middle, and ring fingers (like the Boy Scout salute) measure about 5°. Your closed fist is about 10°, the distance between the tips of your index and little fingers with your hand fully spread is about 15°, and from the tip of your thumb to the tip of your little finger with your hand fully spread is about 25°. The measurements for 1°, 5°, and 10° are remarkably close for most people, but the spread hand measurements vary some from one person to another.

Magnitude – Measuring Brightness

More than two thousand years ago, the first recorded attempt to quantify the brightness of sky objects was undertaken by the Greek astronomer Hipparcos. His scale of measurement varied from first to sixth magnitude. First-magnitude stars were the brightest he could see, while sixth-magnitude were the faintest. As the science of astronomy progressed, the magnitude system was refined to allow precise measurements of all celestial objects. For example, Venus is brighter than the brightest star and reaches a magnitude of more than –4 at times. The brightest star in the sky, Sirius, is magnitude –1. From a dark, clear site, the faintest stars most can see are magnitude 6, just as Hipparcos designated. Telescopes and binoculars collect and concentrate light, allowing us to see fainter objects. Table 2.1 estimates the magnitude limits visible in instruments of various sizes.

A difference of 1 in magnitude is actually a difference of $2\frac{1}{2}$ in brightness. Thus, the difference in brightness between a magnitude-2 star and a magnitude-4 star is a factor of 6.25 (2.5 times 2.5). This explains why larger and larger instruments only gain fractional improvements in limiting magnitude. Nonetheless, these fractional differences are significant. The relatively modest step of just 1 magnitude difference between a 5-inch and an 8-inch telescope brings many thousands of faint objects into view.

We can estimate the magnitude of naked-eye objects using a couple of the most recognizable star formations in the sky. Viewers in the Northern Hemisphere should turn to the

Table 2.1. Magnitude limits with instruments of various sizes

Instrument	Aperture	Limiting magnitude
Naked eye	About 7 mm	6
Binoculars	50 mm	9.5
Telescopes	60 mm (2.4 in)	10.5
	80 mm (3.1 in)	11.3
	4 in	11.7
	114 mm (4.5 in)	12
	5 in	12.3
	8 in	13.3
	11 in	14
	14 in	14.6

Little Dipper. In the Southern Hemisphere, refer to the Southern Cross. The magnitudes of their various stars are shown in Figure 2.3.

Magnitude figures for deep sky objects are not as clear-cut as those for stars and planets. Generally the entire luminosity of the object is "summed up" and reported as if it were a single-point light source (a star). So, a very large object of several arc minutes could be reported with a fairly bright magnitude, but appear very faint in the eyepiece. This is typically the case for nebulae and galaxies. Consider the Andromeda Galaxy, M31. It is generally reported as approximately magnitude 3.4, but that 3.4 is spread out in an area of about 180 by 60 arc minutes. This is about 6 times the width of the full Moon! So, while a star of magnitude 3.4 is easily visible to the naked eye, M31 requires dark clear skies to be glimpsed without optical aid.

In the case of deep sky objects, a better measure of luminosity is "surface brightness". Surface brightness is not standardized and thus varies from one recorder to another, but is generally a measurement of magnitude per square arc minute. Using such a measure we can better compare deep sky objects and determine whether we should be able to view them in our telescope or binoculars. One slight complication for using surface brightness is the fact that not all objects are uniformly bright across their entire surface. Again consider M31. The core is many times brighter than the surrounding spiral arms. Thus, the

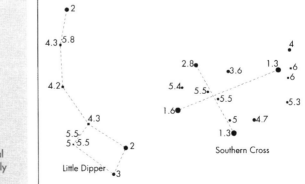

Figure 2.3. Visual magnitudes of easily identifiable stars.

surface brightness of the core is higher than the average surface brightness of the entire galaxy.

Seeing Conditions

Other than the obvious difference between a clear and a cloudy night, most folks don't realize how variable our view of the night sky really is. Some nights you can make out great detail on Jupiter, while other nights you are lucky to see two bands. One night you can see a faint globular cluster quite clearly, other nights it might be invisible. Various factors affect our seeing conditions, but three are most critical: seeing, transparency, and light pollution.

Seeing is mainly our observation of the distortion caused by different layers of air going different directions due to wind, the jet stream, temperature differences, etc. It causes images to waver. Think of this as the hot air wavering above a blacktop road in the summer. In the night sky, it causes the stars to twinkle. In the eyepiece you can most easily see the wavering air when you increase the magnification on a planet or the Moon, but it affects all objects viewed.

Transparency is our observation of the distortion caused by particles in the air. Clouds are obviously the extreme, but dust, smog, moisture, and other particles all limit the transparency of the air. Transparency is the same factor that affects what pilots call visibility – "Today visibility is 10 miles." When observing from a very dark site, poor transparency causes objects to appear less bright. When observing from a site with any appreciable light pollution (and light from the Moon) the effect is many times worse. Those particles reflect light back down to the ground. This results in a "glowing" sky and makes faint objects very difficult or impossible to see. The effect on contrast (difference in light levels) is to make the black velvety sky behind the object not black, but gray. Sky contrast is critical for faint objects, but does not affect contrast much on the surface of bright objects like the Moon and brighter planets.

Light pollution is generally considered to be any man-made light that is directed upward. Houses, cars, storefronts, and streetlights are all prime sources of light pollution. Any level of light pollution will decrease the contrast between celestial objects and the background sky. Naturally the problem is worse around cities, but even small towns have appreciable levels of light pollution. The truth is, there are very few populated places on Earth that still offer truly dark skies. Growing up in rural Indiana, on very dark and clear nights I could actually see the "light dome" above Chicago from more than 70 miles away! Some communities have adopted local regulations to cut down on light pollution. Connecticut has passed a law requiring full-cutoff lights – those that do not send light upward – and Colorado has passed a similar one. Besides spoiling the night sky, all light that goes upward is simply wasted energy.

One of the best ways to gauge the night sky is by determining the faintest stars you are able to see from your site. Using the magnitudes listed in Figure 2.3, you can accurately assess current conditions. Wait for your eyes to become at least initially dark-adapted (20 to 30 minutes), then determine the faintest magnitude visible.

Nights of poor transparency frequently offer steady seeing. Steady seeing is critical for viewing details on the planets and the Moon. Good transparency usually means poor seeing, yet transparency is much more important for deep sky objects. If transparency is just average, even a quarter Moon will wash out most deep sky objects such as galaxies and nebulae. Other than traveling to a dark site, you cannot do much about the general light pollution, skyglow, in your area. But you will want to set up to observe in a location where all local light sources are blocked from your view. For example, moving around the corner of the house might block your neighbor's porch light. Most importantly, get out as often as you can – you never know when that night of perfect seeing conditions will happen.

Observation Technique

While buying a larger telescope is one way to see more difficult astronomical objects, improving your observing technique plays a big role. Experienced observers can always see more in the eyepiece than a newcomer can. Here are some pointers.

First of all, you will most likely need to temper your expectations. Don't expect the view through the eyepiece to match the wonderful photographs you have seen in magazines. Most of those photos are long exposures, sometimes hours long. All that time allows the light to "build up" on the film or CCD chip, providing a bright, colorful image. Your views will be much fainter. Additionally, film and CCD are much more sensitive to color in faint light than your eyes. Other than the planets and a few stars, most objects will be gray in the eyepiece. And finally, things will be much smaller in the eyepiece than they are in an enlarged photograph.

Allow your eyes to become fully dark-adapted and guard that adaptation throughout the night. Everyone is familiar with the fact that after stepping out of a lighted building, it is not possible to see much at all. But within just a few minutes, you are able to see fainter objects (both on the ground and in the sky) than you could initially.

Your night vision improves dramatically in the first 30 minutes of avoiding bright lights. After that, the process slows down as your ability to see fainter light sources continues to improve slightly for a couple of hours. But all it takes is one look into a bright light and your eyes get to start over again. Red light is the least damaging to our night vision, thus astronomers use red flashlights when they need light. But the red light should be very dim as even a bright red light reverses some of your dark adaptation. Also note that the Moon through binoculars or a telescope is an extremely bright object. Leave your lunar viewing for last if possible. If a local light source doesn't allow you to become fully dark-adapted, try draping a dark cloth over your head and the eyepiece.

Use averted vision for faint objects like galaxies, nebulae, and globular clusters. The central area of the retina is best at detecting color while the areas off center are better at detecting faint light. When viewing a faint object, try focusing to the side of the object while concentrating on the object itself. You will generally be able to make out fainter detail in this way. Look to the outside when viewing with one eye (to the right when using your right eye, to the left when using your left eye) or look up when using binoculars. It takes practice, but it is effective.

Move the scope slightly from side to side or tap it. Sometimes an object will be in the eyepiece, but you can't locate it. If you move the scope slightly, the motion will often make the object visible. Once you have located it in this way, you can generally study it with averted vision.

Try different magnifications on an object. Brighter, compact objects hold up well and show more detail with higher magnification. Planets, planetary nebulae, globular clusters, and some small galaxies are in this category. Larger and more diffuse objects generally require low magnification to display the best view. Nebulae, open clusters, and most galaxies fall into this group.

Spend some time on each object. When you first start out in astronomy, the tendency is to jump from object to object, as everything is new to you – resist that urge. What details can you make out on a planet's disk? Moments of clear seeing will reward you with subtle details that are not initially obvious. Can you make out any features in that faint smudge that is a galaxy millions of light years away? How many individual stars can you pick out in that star cluster?

If you wear glasses but do not suffer from astigmatism, feel free to remove your glasses and focus the telescope for a sharp view. The only disadvantages are that anyone else looking through your scope will need to refocus and you might need to use your glasses to view charts and other materials.

Telescopes require time to cool down to the surrounding night air. Until they reach equilibrium, air currents inside the scope will spoil high magnification views. Large, closed-tube telescopes may require more than an hour, while small, open-tube scopes may equalize in a matter of minutes. Some large scopes may not be able to cool fast enough to keep up with quickly dropping nighttime temperatures. To help minimize cool-down time, it is best to store your telescope in a dry, unheated building. If this is not practical, get a head start by setting your scope outside as soon as the Sun goes down, before you intend to use it.

Similarly, heat rising from an asphalt surface or dark-tiled roof will spoil views. This causes the view in the eyepiece to waver and "boil". Consider the surroundings when choosing a location to set up your telescope.

Many amateur astronomers record their observation sessions in a log. In the simplest form, a log is a small notebook where the observer records the date, the objects viewed, and a short description of each. On the other end of the scale are complex databases running on a personal computer that help you to easily search your entries, group together each observation of a particular object, record the eyepiece, magnification, and much more. Most settle on something in between.

A comprehensive approach to recording an observing session includes noting the following in your log:

- Location, date, and time.
- Instrument(s) used and vital statistics about them – aperture, focal length, etc.
- Seeing conditions: seeing, transparency, and level of light pollution.
- Objects and a description of your observations of each.

You certainly will not want to spend much time writing grammatically perfect notes while you are out under the stars, so try this instead. Take a small notepad (or loose paper on a clipboard) out with you and leave your "official" logbook inside. Start by noting the location, date, and time on the notepad. Then simply jot notes about each object. Later that night, or perhaps in the next day or two, refer to your notes to record complete entries in your permanent logbook.

Besides providing a permanent record of your nighttime adventures, a logbook also helps you to see more. In your quest to record what you observe, you concentrate on small details and really see the object. Some observers include sketches of objects in their logbook. More than anything else, drawing will truly help you to see more detail than you thought possible.

Even if you don't continue your log throughout your viewing career, you will find that maintaining a logbook for your first year will really help you to improve in your observation abilities and will build a firm foundation for a lifetime of enjoyment. Plus, the initial wonder of it all will be reflected in your log entries and skimming through your logbook on cloudy nights can recapture that "new" feeling from when you first started out.

Figure 2.4 shows a sample log sheet and sketch sheet. Both can be printed on standard paper and kept in a loose-leaf binder. Visit the Downloads section of my NexStar Resource Site – http://www.NexStarSite.com – to download printable copies.

As you gain more observing experience, you will notice that even your first looks at an object yield greater detail than your painstaking efforts as a beginner. The most important thing is to have fun and learn as you observe. Develop your observation technique and you will develop astronomy into a lifelong interest.

What's Up There?

When I'm out with my telescope and a passerby stops to ask what I'm looking at, the most common question is, "Wha'cha looking at, the Moon?" Well, occasionally that is just what I'm studying, but far more often it is something quite different from our Moon; the sky offers a wonderful variety of objects for our enjoyment. Astronomy is not simply a visual activity, but it is also a mental activity. It is much more enjoyable to understand a little about the objects you are viewing rather than just jumping from object to object in order to simply put another checkmark next to an item on a list.

The Solar System

The planets and other solar system objects can provide hours of enjoyment, both with naked-eye views and views through a telescope. There are nine major planets, countless asteroids and comets, and many more objects revolving around the Sun. Each is held in its orbit by the mutual gravity of the object and the Sun. Some, such as the planets and major asteroids, are visible at frequent intervals of at least every year, while others, such as comets, might only venture close enough to Earth to be seen once every few hundred years.

Each of the planets has a different length of year and their individual motion, coupled with the motion of our Earth, causes them to move against the backdrop of the much more distant stars. For example, while the constellation Orion will always be visible during December during our lifetimes, the planet Jupiter might be high overhead in December one year, but not until January the next, and then not until February the year after that. The closer a planet is to the Sun, the faster it moves – in other words, the shorter its year. You can keep track of the current locations of the planets with the monthly pullout star charts in various magazines or on several Internet web sites; refer to Chapter 6 and Appendix A for suggestions.

The Sun (Figure 2.5) At the heart of our little corner of the universe is the local star, our Sun. Traveling around the Sun in elliptical orbits are the nine planets of our solar system. Besides making summer days almost too hot to bear and turning many of us pink when we don't pay the proper respect, the Sun provides the energy that is the source of most life on the Earth.

While most think of astronomy as a strictly nighttime activity, we can also study the Sun with the proper equipment. But it is extremely dangerous to view the Sun without taking the proper precautions. Refer to Chapter 8 for advice on equipment necessary for solar viewing. After we are properly equipped, we can observe the changing pattern of sunspots as they move across the surface of the Sun. Most of the sunspots we can easily see are several times the diameter of the Earth! Approximately every 11 years the Sun reaches maximum activity with many more sunspots than usual. The most recent solar maximum was 2001–2002.

The Moon (Figure 2.6) The closest major body to the Earth is of course our only natural satellite, the Moon. At about a quarter the diameter of the Earth, the Moon is a cold, barren place. Long ago it would have lost any atmosphere it might have had. Other than very small amounts of ice in craters near its poles, there is no water to speak of. The

Observer Name:	Date:
Site Location:	Time:

Instrument	Conditions (1 – worst, 10 – best)
Name:	Seeing:
Aperture:	Transparency:
Focal Length:	Sky Darkness:
Telescope Type:	Limiting Magnitude:

Observation Notes:

a

Figure 2.4. Sample observation logbook sheets.

Observer Name:	Date:
Site Location:	Time:

Instrument	Conditions (1 – worst, 10 – best)
Name:	Seeing:
Aperture:	Transparency:
Focal Length:	Sky Darkness:
Telescope Type:	Limiting Magnitude:
Eyepiece(s):	° above Horizon:
Magnification(s):	
Filter:	Field Diameter:

Object	Constellation	Coordinates
		RA h m
		Dec

N

b

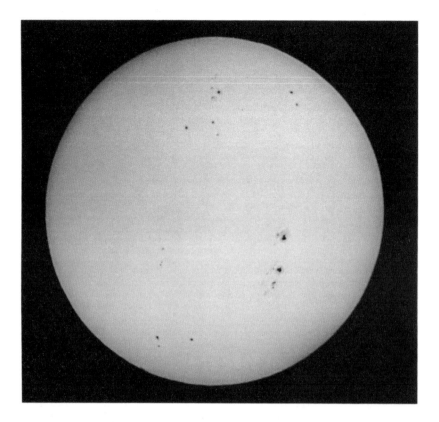

Figure 2.5. The Sun imaged with a NexStar 5 and a digital camera. Image by Dr. Mario Mariani.

craters covering its surface are a silent reminder of the violent past (and occasional reoccurrences) of our solar system – they were all created as objects impacted the lunar surface.

But it is those amazing surface features that fascinate us when viewed with even the slightest magnification. It is easy to get lost among the various craters, mountains, valleys, fissures, and other geological formations. As the Moon makes it way around the Earth about once every 29 days, it presents us with differing phases as the angle between us, the Moon and the Sun varies. Contrary to popular belief, the best time to view the Moon is not when it is full. Rather, at any other time, the greatest detail can be seen along the terminator – the line between the sunlit and dark sides of the Moon. The surface of the Moon at the terminator is experiencing sunrise or sunset and the low angle of the Sun produces long shadows, just as it does here on Earth. Those long shadows make surface features more pronounced and much easier to see.

As your early observations of the Moon will show – it's bright! One accessory that is a necessity for every amateur astronomer is a Moon filter, as discussed in Chapter 8. This cuts down most of the glare and makes viewing the Moon much more comfortable. Even

Figure 2.6. The Moon imaged with a NexStar 11 GPS and a digital camera. Image by Christopher Engell.

with a Moon filter you should generally do your lunar observing last on any night as viewing the Moon will ruin your eyes' dark adaptation.

Also, when the Moon is more than just a sliver in the sky, it tends to wash out the fainter deep sky objects. Thus, your observations of galaxies, nebulae, and such will generally need to wait until the Moon sets or you should hunt for them before the Moon rises. During nights near the full Moon, only the very brightest deep sky objects will provide satisfying views in the eyepiece.

Mercury The planet with the closest orbit to the Sun is Mercury. Mercury is so close to the Sun that we seldom see it and when we do it is just after sunset or just before sunrise. Due to the angle between the Earth, Mercury and the Sun, Mercury exhibits phases when viewed through a telescope as we can see both the side illuminated by the Sun and the side in the shadow of Mercury's night. Other than these phases, you will see no detail when viewing Mercury.

Venus Next in order from the Sun is the planet Venus. Venus is the brightest nighttime object in the sky, after the Moon. The dazzling brightness of the planet causes some to mistake it for aircraft lights! Venus is also relatively close to the Sun, visible either in the few hours after sunset or before sunrise. When Venus is out after sunset, it is often called the "Evening Star" (although we amateur astronomers know it is a planet, not a star!) as it is the first "star" to show as the sky darkens through twilight. When visible before sunrise,

Figure 2.7. Mars imaged with a NexStar 5 and a digital camera. Image by Dr. Mario Mariani.

Venus becomes the "Morning Star". The stunning brightness is largely due to the light-colored clouds that continually cover the planet. For the same reasons as Mercury, in a telescope Venus displays phases as it travels around the Sun. There is no real detail to be seen viewing Venus other than the changing phases.

Most find Venus to be simply a beautiful sight with the naked eye, since there really isn't much to see in the telescope. Occasionally Venus, or any of the other planets for that matter, will pass in front of – occult – a background star. Or, rarely, the Moon will occult one of the planets. These are occasions to train our telescopes on Venus, rare opportunities not to be missed.

On 8 June 2004, a very rare event including Venus will present itself to observers from Guam through Asia over to Europe. Venus will transit (pass in front of) the Sun. A telescope equipped with a proper solar filter (see Chapter 8) will view an occurrence that won't be seen again from Earth until 2012.

Mars   (Figure 2.7) The planet occupying the orbit outside of Earth's is the red planet, Mars. A bit smaller than Earth, Mars is a rocky, dusty planet. Robot missions to Mars have shown evidence that in the distant past water flowed on the planet's surface. While none exists there now, astronomers have recently found evidence of water ice below the surface, as well as that found in the polar icecaps.

Viewing Mars is best when it is at opposition – when Mars and the Earth are on the same side of the Sun, with all three forming a straight line. Oppositions occur about every 2 years and some oppositions are much better than others. About every 15 or 16 years, the elliptical orbits of Earth and Mars cooperate to bring us much closer to Mars than during other oppositions. The cycle is not exact by human standards, so it is best to keep up to date with information in monthly magazines, newsletters, and web sites. The opposition in August 2003 is one of those best passes, although 2005 will also provide better than average views.

Viewing Mars in a telescope should present a pink to red disk. Look for surface features such as the light-colored polar caps and various dark surface markings. The angle of Mars will not always present a good view of the caps, and generally only one of them will be visible. Watch for changes in the size of the caps as Mars moves through its seasons. Very good seeing conditions will be required to see the caps in smaller scopes.

The dark surface features also change with time. Dust storms on Mars are usually the cause as fierce winds shift the lighter-colored dust over the darker-colored bedrock. Try

Figure 2.8. Jupiter imaged with a NexStar 8 and a digital camera. Image by Al Fugiero.

your hand at sketching Mars on as many nights as possible and compare your drawings with a map of the Martian surface. You might find that some of the dark patches you draw are markedly different over a month.

Asteroids Asteroids or, as they are sometimes called, minor planets are distributed in various places throughout the solar system. There are many in Jupiter's orbit in two clusters in front of and behind Jupiter. There are some in errant orbits that occasionally bring them too close to Earth for comfort. There are asteroids beyond count in the Kuiper Belt, a donut-shaped region beyond the orbit of Neptune. But the ones of most interest to amateur astronomers are located in an orbit between Mars and Jupiter – an area known as the Asteroid Belt. Several of these asteroids are within reach of backyard telescopes, although just as points of light.

Jupiter (Figure 2.8) King of the planets, Jupiter is larger in mass than all of the other eight planets combined. Jupiter is a large ball consisting mainly of gases. Jupiter, Saturn, Uranus, and Neptune are similar in nature and are known as the gas giants. And giant Jupiter is, even from our vantage point here on Earth. Only Venus, due to its much closer proximity, occasionally presents a larger disk in the eyepiece of a telescope. And only Venus is brighter than Jupiter.

But none of the other planets compares to the view of Jupiter in a telescope. Jupiter shows detail in small telescopes and even the smallest optical aid will show its four bright moons. In small telescopes you can make out two or three of the darkest cloud bands, and as the scope gets bigger, the more you will see. 4-inch scopes can see multiple bands and the Great Red Spot. Larger scopes can see details in the bands such as texture, loops, and ovals, often in vivid color. Also visible in larger scopes are transits of the moons across Jupiter's surface as well as the inky black dots of the moons' shadows as they transit the planet surface.

Jupiter spins at an incredible rate of one rotation in less than 10 hours, and thus the view is changing continuously throughout the night. It is common for an observer to revisit Jupiter many times during a long night session. Try your hand at sketching Jupiter one night – the concentration will bring out more detail than just a casual look. You will soon know why it is called the "amateur's planet".

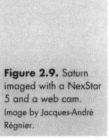

Figure 2.9. Saturn imaged with a NexStar 5 and a web cam. Image by Jacques-André Régnier.

Saturn (Figure 2.9) While Jupiter provides the most detail in the eyepiece of amateur astronomy equipment, Saturn is the most dazzling sight. Even small telescopes will show Saturn's rings, eliciting a "wow" from almost every first-time viewer. Moderate-sized scopes will show the dark void between the "A" and "B" rings – the Cassini Division. Larger telescopes will show a fainter "C" ring, also known as the Crepe ring. As the Earth and Saturn make their way around the Sun, periodically the Earth passes through the plane of the rings and they are not visible. At this time we are viewing these very thin structures edge-on.

Beyond simply seeing the rings, look for the rings' shadow on the planet surface. Also, look for the planet's shadow on the part of the rings behind the planet. In a larger scope, you can make out faint detail on the surface of the planet. Saturn, like Jupiter, has cloud bands in its upper atmosphere. The planet itself is also flattened prominently owing to Saturn's low density and high rate of spin. This is easily observed in a telescope.

Saturn has a dozen named moons and many more unnamed. The largest and brightest, Titan, is visible in most telescopes. Larger telescopes will show at least five of Saturn's moons.

Uranus, Neptune, and Pluto Seventh from the Sun is the planet Uranus. While the other five planets discussed thus far are visible to the naked eye, optical aid is required to sight Uranus. In fact, this is the first planet to have been discovered after the invention of the telescope. Although it can be found in small telescopes, it was not until 1781 that William Herschel discovered Uranus. A small telescope will see it as a pale disk; a larger telescope will show a fairly bright disk, blue in color. Uranus has faint rings, but they cannot be seen in Earth-based telescopes.

Neptune is the eighth planet from the Sun – at least, most of the time. Pluto's orbit crosses inside of Neptune's during 20 years of its 249-year trip around the Sun. In a telescope, Neptune presents a disk that is easily distinguished from background stars. Pluto is hard to make out in anything less than a 6-inch scope. Even then, you will need a good star atlas to help you discern the faint, star-like point of light that is Pluto.

Comets Comets are the true wanderers of our solar system – icy bodies in wildly elliptical orbits that plunge towards the Sun and are whipped back out to the far reaches of the

Table 2.2. Annual meteor showers

Shower	Date	Radiant
Quadrantid	3/4 January	Draco
Lyrid	21/22 April	Lyra
Perseid	12/13 August	Perseus
Leonid	18/19 November	Leo
Geminid	13/14 December	Gemini

solar system. Not all comets are so lucky; some are vaporized by hurtling directly into the Sun or by simply passing too close. Most of the comets in our solar system are long-period comets that are from the Oort Cloud, a region far from the Sun containing perhaps millions of potential comets. Short-period comets come from the Kuiper Belt. Pluto and its moon Charon, while certainly not comets, are the most prominent residents of the Kuiper Belt region.

As a comet approaches the Sun, the solar wind – charged particles and radiation speeding away from the Sun – creates a tail (sometimes more than one!) of glowing gases extending from the head of the comet. This tail always points away from the Sun in response to the solar wind. Some comets have tails extending millions of kilometers into space. Occasionally a comet comes along that is readily visible to the naked eye, but most comets require the use of binoculars or a telescope to be seen.

Meteors and Meteorites
When objects enter the Earth's atmosphere, friction causes them to burn brightly. These streaks of light, sometimes called shooting stars, are known as meteors. Most of these objects are tiny, just grains of sand, and burn completely in the upper atmosphere. Occasionally a much larger object enters the atmosphere and survives all the way to the surface. These survivors are known as meteorites.

From a truly dark site, you can observe several meteors every night. On certain nights each year we experience meteor showers, when the frequency may climb to hundreds per hour. Rarely one of these annual showers becomes a meteor storm with hundreds of meteors visible per minute. These annual showers are the results of the Earth passing through the trail of debris left behind by a comet. Table 2.2 presents some of the best annual meteor showers.

The dates given are approximate and vary from year to year. Also, the days before and after a peak usually experience increased meteor activity. The radiant is the constellation that the majority of the meteors seem to trace back to.

To view meteors you do not need any optical gear, just your eyes and a clear, dark sky. Most meteors are seen in the early morning hours after about 1 a.m., as your location on Earth is headed into the wind, so to speak, at that time. Setup a reclining lawn chair or simply lie on a blanket on the ground and look in the general direction of the radiant. Remember to dress warmly; it can be quite chilly before dawn any time of year.

Manmade Satellites
Since the first satellite, Sputnik, was launched in 1957, we have put thousands of objects in orbit around the Earth. Many of these are visible to the naked eye just after sunset. Watch for a bright object traveling briskly along until it suddenly snaps off, as if someone had turned off a light switch. Most likely you will have just seen a communications satellite. One particular type of communications satellite, Iridium, produces a brilliant, short-duration flash as sunlight glints off its mirror-like surface. If a light trail is especially bright and travels across much of the sky, you likely got a firsthand

view of the International Space Station (ISS). Don't be surprised if the streaking light of a satellite moves briskly through the field of view of your telescope some night. Also, there are many pieces of junk in orbit: spent rocket boosters, lost communications satellites and such, that flash on and off sporadically as they tumble through space.

Deep Sky Objects – DSOs

Objects outside of our solar system are known collectively as deep sky objects or DSOs. DSOs range from individual stars to cities of stars known as galaxies. Our own galaxy, the Milky Way, with its billions of stars, is but one of the billions of galaxies in the universe. While the distances in our one solar system are immense – consider that it takes more than a year for our current spacecraft to travel from the Earth to Mars – the distances we encounter when considering DSOs are almost unfathomable. When dealing with such distances our Earth-bound units of kilometers and miles fail us. Instead astronomers generally rely on a unit known as the light-year.

Deep Sky Object Catalogs

Most items in the sky are listed in one or more of the various "catalogs" that astronomers have created over the years. For example, the Andromeda Galaxy is known as M31 and NGC224. Some of the most useful catalogs for amateur astronomers are:

- Bayer designations for stars – a system of designating the brightest stars in each constellation by Greek alphabet – for example the brightest star in Ursa Major (Dubhe) is known as α (alpha) Ursa Majoris.

- Flamsteed designation for stars – a system of designating the stars in each constellation by number – thus Dubhe is also known as 1 Ursa Majoris.

- Various other star catalogs list stars as serialized numbers scattered all around the sky – Smithsonian Astrophysical Observatory (SAO), Henry Draper (HD), Hipparcos (HIP), and Hubble Guide Star (GSC) are common star catalogs.

- Messier Catalog – 110 DSOs of various types of objects. Most of the brighter deep sky objects visible in the Northern Hemisphere are listed in Charles Messier's catalog. Generally these objects are the first DSO targets for beginning amateur astronomers. Messier 1 would be commonly referred to as "M1".

- Caldwell Catalog – 109 DSOs to expand beyond the Messier list. Many of these objects are only visible to observers in the Southern Hemisphere.

- New General Catalog (NGC) – 7840 objects and Index Catalog (IC) – 5386 objects – a large collection of DSOs cataloged by J.L.E. Dreyer. Other than individual and double stars, the majority of DSOs of interest to amateur astronomers are found in these two catalogs.

Other catalogs will be of use to you after you gain experience, but these will provide many targets for your beginning years as an amateur astronomer.

Light travels at a constant speed of approximately 186 000 miles per second (300 000 kilometers per second). A light-year is the distance light travels in one year – about 5.9 trillion miles or 9.5 trillion kilometers. Thus, a light-year is not a measure of time, but rather of incredible distance. The closest Sun-like (main sequence) star is Alpha Centauri at a distance of 4.3 light years. Thus, the light we see from Alpha Centauri has been traveling for a little more than 4 years or a distance of about 25.4 trillion miles (42.2 trillion kilometers).

Galaxies are **much** farther away. The nearest major galaxy similar to our Milky Way Galaxy is the Andromeda Galaxy. Current calculations put it about 2.2 **million** light-years away. You can do the math if you are interested in miles or kilometers, but consider that the light from the Andromeda Galaxy has been traveling for 2.2 million years before it reaches your eyes.

DSOs are the favorite targets of many observers, but most of them are very faint, resulting in their common nickname – faint fuzzies. Many of them require fairly large telescopes to show much detail or even to be detected in the eyepiece. Unlike the planets and the Moon, DSOs are easily washed out by light pollution, making them more difficult targets for the observer in the city. DSOs are a marvelously varied lot though, and a good number of them are readily visible even in smaller scopes.

Stars, Variable Stars, and Double Stars

Even in the largest telescopes individual stars never present a disk like a planet. They are just tiny points of light, but all deep sky objects are visible to us due to the light and radiation produced in the nuclear furnaces of individual stars. Some stars are interesting in their own right, due to their vivid color. For example, Betelgeuse, the upper left star in the hourglass shape of the constellation Orion, is a brilliant red. Beyond that, the average individual star provides little to hold our interest. All of the stars visible to us as individual points of light are inhabitants of our own galaxy.

Variable stars fluctuate in brightness, sometimes in a regular cycle, sometimes chaotically. Some amateur astronomers enjoy recording these variations, an interest that requires regular observations and a good eye for detail.

Double stars are perhaps the most interesting type of star for amateur astronomers. Double stars, when visible to the naked eye, generally appear to be a single star. But when we magnify them with binoculars or a telescope, we can "split" them into two or more individual stars. Sometimes these stars are binary or multiple star systems – stars that revolve around each other and travel through space as a group. Other double stars are simply optical doubles – they may be hundreds of light-years apart but their chance alignment from our vantage point on Earth provides us with two closely placed stars. The closer those stars appear, the larger is the telescope required to split them.

Open Clusters

Open star clusters are groups of stars that were born together from the same cloud of gas. They are generally young, bright stars that are slowly drifting apart, but at a rate that won't spoil our view of the cluster for hundreds of thousands or even millions of years. Many of the stars in the sky started their lives in a cluster, but have since drifted to their current locations after billions of years.

The open clusters we can readily observe are relatively close to Earth – almost all of them are less than 10 000 light-years distant. Smaller, wide-field telescopes provide some of the best views of many of the open clusters, particularly those closest to us. A larger scope will help to bring out the fainter members of the cluster, especially in viewing from a light-polluted site.

Globular Clusters

(Figure 2.10) Globular star clusters are immense balls of stars held tightly together by their mutual gravity. All of the known globular clusters in our galaxy are very old, nearly as old as the galaxy itself, but astronomers have discovered younger globular clusters in some neighboring galaxies. Some of the largest globular clusters contain more than a million stars packed into an area of a few hundred light-years across – imagine the night sky from the middle of such a cluster!

More than any other deep sky object, globular clusters produce their best views at high magnifications. A dark site, a large telescope, and a clear night can allow you to "pump up the power" and observe a nearly three-dimensional view of these jewels of the night sky.

Figure 2.10. Omega Centauri globular cluster imaged with a NexStar 5 and a digital camera. Image by Dr. Mario Mariani.

Larger telescopes will easily resolve stars from the general glow of the cluster, particularly if you use averted vision. The challenge with smaller scopes might be to simply detect the faint glow in the eyepiece.

Nebulae

Nebulae (the plural form of nebula) are a general category of DSOs that are faint, diffuse and, well, nebulous in appearance. A hundred years ago, when telescope quality did not provide the highly resolved views we enjoy today, almost all DSOs were referred to as nebulae. Today we use the term to refer to clouds of gas and dust.

In most general terms we can further break them down into bright and dark nebula. Bright nebulae are powered by stars embedded in them. Some are reflection nebulae – the dust and gas simply reflect the light of the nearby stars. Others are emission nebulae – gas in the nebula emits its own light due to the molecules of the gas being energized by radiation from the nearby stars. Dark nebulae are visible as dark patches where dust is blocking the light of background stars.

Most nebulae are very difficult to see without truly dark and clear skies. Take advantage of any trips to dark observation sites by preparing a list of nebulae. Look for details in the clouds of gas such as texture, lighter and darker structure (especially dark lanes), and background or embedded stars.

Planetary Nebulae

(Figure 2.11) Another type of faint fuzzy is the planetary nebula. The term planetary nebula came from some of the first astronomers with access to telescopes. The brighter planetary nebulae they were able to detect were similar to planets in the size and appearance presented in the eyepiece, yet they were obviously gaseous in nature.

Planetary nebulae are the remains of old stars similar in size to our Sun. In fact, in a few billion years, if everything goes as expected, our Sun will become a planetary nebula. Due to changes in the nuclear reaction occurring in the star's core, it eventually becomes unstable and an explosive reaction blows off most of the gases in the outer portion of the star. The result is a small, hot, white dwarf star with an expanding cloud of gas escaping away at high speed. Radiation from the star energizes the gas, causing it to glow.

Figure 2.11. The Ring Nebula imaged with a NexStar 11 GPS and a CCD camera. Image by Vernon Riley.

Planetary nebulae come in a variety of shapes. Twin cones expanding out from the central star are well known. So are spheres of glowing gas around the parent star. Others are more intricate shapes, suggesting very complex circumstances at the time of the explosion or perhaps that the star had a binary companion that helped create a convoluted shape due to its gravity.

There are many planetary nebulae visible to amateur astronomers. A few even display noticeable color. Use higher magnification to see if you can detect additional detail.

Similar to planetary nebulae are supernova remnants. Occasionally a star really explodes resulting in an elaborate cloud of hydrogen, helium, and heavier elements. The supernova itself was extremely bright. Nearby supernovae would be visible even in the light of day. The remaining cloud of gas is usually very faint and difficult to see in a backyard telescope.

Galaxies (Figure 2.12) Containing as many as billions of stars, galaxies come in a wide variety of shapes and sizes. Spiral galaxies like our own Milky Way Galaxy are among the largest and most common. With their large central bulge of stars and graceful sweeping arms, spiral galaxies are one of the most majestic sights you will encounter in the sky. Giant elliptical galaxies are incredibly large galaxies with many billions of stars amassed in an egg-like shape. Smaller dwarf galaxies are often classified as irregular galaxies – millions of stars seemingly dropped haphazardly like a child's jacks against the backdrop of the sky.

These huge collections of stars are visible from unbelievable distances. As noted earlier, the closest galaxy to us is an incredible 2.2 million light-years distant. Other galaxies

Figure 2.12. The Whirlpool Galaxy imaged with a NexStar 11 GPS and a CCD camera. Image by Vernon Riley.

visible to the amateur astronomer are more than 60 million light-years from the Earth. The light we see today left some of these galaxies when dinosaurs still ruled the Earth!

Due to the extreme distances, most galaxies are very faint. While you might detect them with a large scope in the city, the views improve enormously under dark skies. You will also have a much better view when the sky is very transparent. Under such clear, dark skies a larger telescope will show the arms of spiral galaxies, dust lanes that obscure the light from the billions of stars, and other subtle details.

Incredibly large though galaxies are, they are not the largest structures in the universe. Typically, small numbers of galaxies are held together by their mutual gravity in a structure astronomers call galaxy clusters. Our Milky Way Galaxy is one of about half a dozen galaxies in the cluster known as the Local Group. Large backyard telescopes can show us other galaxy clusters scattered around the sky.

Furthermore, the billions of galaxies in the universe are organized into superclusters – tens of thousands of galaxies arranged in waves and knots throughout the immense expanses of space. Superclusters are not visible directly, but rather, by mapping the location of millions of galaxies, astronomers have created a three-dimensional model that reveals their existence.

Equipment Basics

Never before has such a wide variety of quality equipment been available to the amateur astronomer. With hundreds of models of telescopes to choose from, we need a basic understanding of the most common designs to make a wise purchase. Although there are so many models to choose from, as we will see there are only a few common designs on the market.

Basic Terms

Before discussing equipment, there are a few terms we must understand.

Objective This is the component of the optical system that collects the light from the sky, allowing us to see all those faint objects that our eye cannot detect on its own. In binoculars and some telescopes this is the lens at the front of the optical tube. In other telescopes, it is a large, dish-shaped mirror at the back of the optical tube. We measure the diameter (distance across) of the objective and express this as the aperture of the instrument. Generally speaking, the larger the aperture, the more light the instrument collects, and thus the more objects we can see. When we say the NexStar 8 is an 8-inch scope, we are referring to the diameter of the objective.

The relationship between aperture and light-gathering power is geometric, based on the area of the objective, not the diameter. When comparing two scopes, simply square the diameter of their objectives and divide. For example, to compare 80mm and 60mm telescopes, we calculate 80×80 divided by 60×60. The resulting answer of 1.78 indicates that an 80mm objective collects 78% more light than a 60mm objective.

Additionally, larger apertures are capable of resolving finer detail in objects. Thus, a larger aperture can usually split closer double stars, show more detail on the Moon and planets, and show more detail in deep sky objects. More so than light-gathering power, resolution is greatly affected by the overall quality of the instrument's optics.

There is a saying among astronomers that "aperture rules". While this is certainly true, larger aperture comes at a price, both monetarily and practically. Larger-aperture telescopes are more expensive then their smaller siblings. Also, larger-aperture telescopes require much more effort to transport, set up, and store. Consider carefully the size of telescope you will purchase. The best telescope for you is not necessarily the most expensive model you can buy. It is the one that you will use the most. An exquisite 14-inch telescope will provide incredible views, but if you only find the energy to set it up once every couple of months, perhaps you would more enjoy a 5-inch scope that takes just minutes to carry outdoors and put into use.

Focal Length While some optical designs do not lend themselves to this straight-forward definition, focal length is the distance from the objective to the point where the image comes to focus. Each eyepiece has a measured focal length that is marked on the side of the eyepiece itself. The focal length of a telescope is sometimes found on a label on the optical tube, otherwise you must refer to the manual. Generally speaking, longer focal lengths produce better high-magnification views of the Moon, planets and smaller deep sky objects, while shorter focal lengths allow for wide-field views of large objects like open clusters and large nebulae.

Focal Ratio Focal ratio is the focal length of an objective divided by the diameter of the objective, and is expressed as f/number. For example, an 80mm telescope with a 400mm focal length has a focal ratio of f/5 (400/80 = 5). Smaller numbers are said to be "faster" as they allow for shorter exposures when taking photographs, but it is a mistaken belief that faster focal ratios produce brighter views in the eyepiece. Only more aperture can produce brighter visual images.

Magnification in a Telescope

To calculate the magnification given by a telescope, divide the focal length of the telescope by the focal length of the eyepiece in use. For example, a 1000mm telescope provides a magnification of 100 when we use a 10mm eyepiece (1000/10 = 100). By changing to smaller focal length eyepieces, we increase the magnification. There is an accepted limit to the amount of magnification possible with a telescope. As a general rule, 2 times the aperture in millimeters or 50 times the aperture in inches is a good rule of thumb, although seeing conditions rarely allow us to exceed a magnification of 300. On most nights a magnification limit of 200 or less is more practical.

Beware of purchasing a telescope that is advertised according to its maximum magnification. Many department stores offer a "575× Astronomical Telescope". Upon closer inspection we find that the aperture of the scope is 60 mm, capable of a usable magnification of about 120×. Since magnification is varied simply by using an eyepiece of different focal length, magnification available with the eyepieces provided is one of the least important factors when considering a telescope. Focus on aperture, smooth movement when pointing the scope, a sturdy mount and tripod, and general overall quality of construction. When possible, test the telescope out under the night sky. Stars should focus easily to sharp points of light. Center a bright star and defocus the image, both inside and outside of focus. The resulting bull's-eye rings should appear nearly the same inside and outside of focus. Judging optics requires a bit of experience – something as simple as poor collimation (discussed in Chapter 9) or tube currents in an uncooled scope can make the best optics look terrible. But even a beginner can detect truly poor optical quality.

Lens Coatings

As light reaches a piece of glass, some of it is reflected rather than passing through. If that glass is a lens in our telescope, the reflected light never makes it to our eyes and thus the image loses some of its brightness. If a lens surface is treated with a special anti-reflective (AR) coating this reduces the reflection and transmits more of the light. Many optical components are constructed of multiple lenses. For example, some eyepiece designs have as many as eight lenses! To be fully effective, all air-to-glass surfaces must be coated. Light transmission can be improved with multiple layers of AR coatings.

Manufacturers use the term "coated" to designate that at least one of the air-to-glass surfaces, usually the side of the lens you can see, has been coated with a single layer of AR coating. The next better process is "fully coated" – all air-to-glass surfaces have been treated with AR coatings. "Multi-coated" optics feature at least one surface with multiple layers of AR coatings, while "fully multi-coated" indicates that all air-to-glass surfaces have been treated with multiple layers of AR coatings.

Eye Relief

When you view through a telescope or a pair of binoculars, eye relief is the maximum distance you can position your eye away from the eyepiece and still take in the entire field of view. Less than 5 mm is very uncomfortable for most people and if you wear glasses to view (necessary if you have astigmatism), eye relief of about 20 mm is required. Eye relief is a characteristic of each eyepiece or binocular design.

Exit Pupil

Exit pupil is a measurement of the fully illuminated circle of light that comes from the eyepiece of a telescope or binoculars. To calculate the exit pupil in millimeters, divide the aperture of the objective (in millimeters) by the magnification of the view. For example, if a pair of binoculars has an aperture of 50 mm and a magnification of 10 (10×50 binoculars), they produce a 5mm exit pupil. For a telescope it varies each time we change eyepieces and a simpler formula is to divide the focal length of the eyepiece by the telescope's focal ratio. For example, if you use a 30mm eyepiece in an f/10 telescope, the result is a 3mm exit pupil.

With binoculars, it is best to use an exit pupil that does not exceed the size of your fully dilated pupil. Most observers under the age of 30 have a maximum pupil dilation of about 7 mm under very dark skies. As we age the muscles in our irises stiffen and it is common for older observers to have a maximum dilation of about 5 mm. With a telescope, we can experiment with various eyepieces to produce the best magnification and exit pupil for the object at hand.

Binoculars

How would you like two telescopes, one for each eye? In essence, that is what a pair of binoculars gives you. Our brain and optical system evolved to integrate input from both eyes, so it is not surprising that binoculars seem very natural when used. Binoculars feature a right-side-up, left-to-right-correct image and a wide field of view that makes them very easy to point at your target. Almost anyone can pick up a pair of binoculars and successfully point them at the Moon; the same cannot be said of a telescope.

Binoculars are labeled with designations like 7×35 and 10×50. The first number denotes the magnification the binoculars offer, while the second number is the diameter in millimeters of the objective (front) lenses. As mentioned earlier, the diameter of the objective is also known as the aperture of an instrument. The larger the aperture, the more light the instrument collects and the fainter the objects you will be able to see.

Higher magnifications do allow you to see smaller details in an object, but this comes at a cost. First, it is generally true that as the magnification increases, the field of view decreases. For example, a pair of 10×50 binoculars might show 5° of sky, while a pair 20×70 binoculars might only show 3°. For general sky "sweeping", a wider field of view is highly desirable.

Second, higher magnifications also magnify every little movement of your arms as you try to steady the binoculars. Most individuals cannot successfully steady binoculars higher than 10× when holding by hand. Higher magnifications require a mount, such as a tripod, to steady the view. Image-stabilizing models are the exception. Equipped with computer-controlled prisms that move with every shake of your hand, they provide a steady view at high magnifications. Of course, image-stabilizing binoculars are more expensive than their traditional counterparts.

Although almost any binoculars will allow us to see many more stars as well as some of the brighter deep sky objects, the best general-purpose binoculars for astronomy feature a minimum aperture of 50 mm and magnifications between 7 and 10 times. 7×50 and 10×50 models are quite popular and very reasonably priced. 7×50 produces an exit pupil of about 7 mm, fine for younger eyes under dark skies, while 10×50 produces an exit pupil of 5 mm, more useful in urban settings or for older observers.

When deciding on binoculars, consider the instrument's anti-reflective coatings, fully multi-coated being most desirable. If possible, take them out under the night sky and look at the Moon. Cheaper models will typically exhibit false color around the edge of the Moon while better models will have less color, leading to shaper images. Also, look at a sign off in the distance. Note if the words at the edge of the field of view are as sharply focused as those in the center. Don't panic if they are not; most binoculars do not produce sharply focused images out towards the edge of the field, but there are notable differences and the better a binocular performs in this respect, the more you will enjoy the view. Finally, consider the weight. Very heavy binoculars will not be comfortable for hand-held viewing. Comfort when using binoculars will generally require a reclined positioned; extended periods of looking up can definitely be a pain in the neck! A reclining lawn chair is an excellent solution, particularly if the arms of the chair are suitably positioned to support your arms as you hold the binoculars. A camping mattress or pad can also provide excellent portable comfort for binocular use.

Even after you own a telescope, a good pair of binoculars is a must. They are quick to set up (just take them out of the case) and they are the ultimate in portability. The wide field of view they afford is wonderful for many extended objects and provides the best views of the star fields of our own Milky Way Galaxy. Binoculars are an essential piece of gear for the amateur astronomer.

Telescopes

Although some amateur astronomers start their observing career with binoculars, it isn't long before they are ready to move on to a telescope.

Telescopes differ greatly from binoculars. Telescopes have a much narrower field of view, especially at higher magnifications. As mentioned earlier, while a pair of binoculars generally has a fixed magnification, we can change the magnification offered in a telescope by simply changing the eyepiece. The orientation of the view provided by a telescope is also different. Almost all astronomical telescopes present a mirror image view (backwards from left to right), upside-down view, or both! While this makes little difference when viewing celestial objects (in space there is no "up"), it presents unique challenges when we attempt to match the view in the eyepiece to a diagram or star chart. And perhaps most importantly, a telescope requires a stable mount that makes easy work of tracking objects as they drift across the sky.

The higher cost and added complexity makes the purchase decision much more difficult. Advertisements and store displays don't provide much help, nor do the wonderful color pictures that adorn the packaging of most telescopes. But the situation is not as confusing as it first seems; most telescopes sold are of one of four designs: refractor, Newtonian reflector, Schmidt–Cassegrain, or Maksutov–Cassegrain.

Refractors   (Figure 2.13) Refractors match most people's image of what a telescope should look like – a long tube with a lens at the front and an eyepiece at the back. The lens at the front of the telescope is the objective in a refractor scope. The lens collects the light and focuses it towards the back where the eyepiece magnifies the view. The first telescopes designed, at least as early as the seventeenth century, were refractors.

While high-quality refractors are prized for their crisp, clear views of the Moon and planets, they are not ideal for deep sky objects. The high cost of creating large objective lenses limits the most commonly available refractors to about 6 inches. The most popular high-quality refractors are in the 4-inch range, relatively limited in their ability to detect faint, deep sky objects.

One of the biggest drawbacks to the refractor design is the lens itself. A single lens tends to act like a prism, splitting white light into a rainbow of colors, an effect known as chromatic aberration. On bright objects such as the Moon and planets this can be quite visible. To combat this effect, achromatic designs use two lenses of different shape and glass composition to reduce chromatic aberration. This can be very effective, particularly if the refractor has a long focal length. But it is the long focal length of traditional refractors,

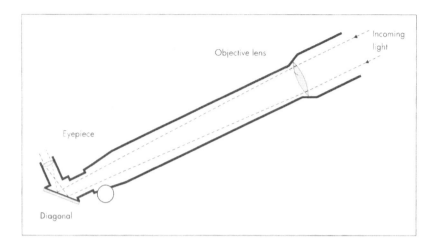

Figure 2.13. The refractor telescope design.

coupled with the expense of creating large-aperture lenses, that caused the refractor to fall in popularity in the 1950s and 1960s.

Refractors saw a great resurgence in the 1990s as short-tube refractors became widely available. Short-tube achromatic designs are quite popular but do exhibit levels of chromatic aberration that some find distracting. Newer, wide-field apochromatic (color-free) designs are very popular and widely used. Apochromatic refractors often sport as many as four lens elements in the objective and provide excellent views of brighter objects. The disadvantage is a price tag generally three to six times higher than achromatic models of the same aperture.

Refractors are equipped with a diagonal at the eyepiece end to reflect the light at a 90° angle in order to provide a more comfortable viewing position. The diagonal is inserted into the focusing tube and the eyepiece is inserted into the diagonal. Without the diagonal, we would find ourselves contorting into various uncomfortable positions to position ourselves near the eyepiece.

Perhaps the greatest strengths of the refractor design are their durability and maintenance-free operation. With reasonable care, a refractor will provide the same high-quality views for many years. The optical components are permanently collimated (aligned) and require no fine-tuning to produce the best image possible. If your time under the stars is limited, or you find you are not mechanically inclined, a refractor might be best for you.

Newtonian Reflectors (Figure 2.14) To combat the effects of chromatic aberration found in refractor telescopes, Sir Isaac Newton created an optical design using two mirrors, a design that now bears his name. In a Newtonian reflector, a dish-shaped mirror is placed at the back of the optical tube. Incoming light reflects off the primary mirror and is focused back towards a small, flat secondary mirror that diverts the light through the side of the tube towards the eyepiece. To create a correctly focused image, the primary mirror should be parabolic in shape. Creating a parabolic mirror is more expensive than creating a simple spherical mirror, so cheaper Newtonians often utilize the lower-cost mirror at the expense of image quality.

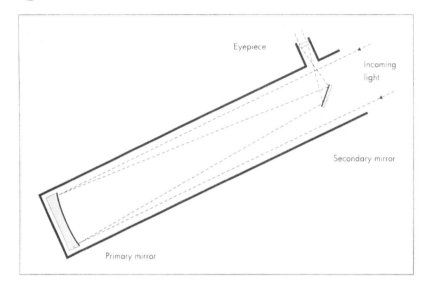

Figure 2.14. The Newtonian reflector telescope design

Besides virtually eliminating chromatic aberration, Newtonians also prove to be the least expensive design per inch of aperture. Virtually all really large amateur scopes, 18-inch and larger, are Newtonian reflectors. In terms of "bang for the buck", nothing beats a Newt.

Newtonians do have some drawbacks. One is the open-tube design. In use, a Newtonian exposes the surface of its mirrors to the elements. Unlike the mirror in your bathroom, a telescope mirror has the reflective coating on the front of the glass. This coating is typically aluminum, covered with a thin protective layer to retard corrosion. But eventually the elements win out and the mirror becomes tarnished. At this time it becomes necessary to have the mirror professionally recoated.

Larger-aperture Newtonians can be very long and heavy. Truly large Newts usually sport a frame of thin tubes to support the primary and secondary mirrors as well as the eyepiece holder and focuser. This frame can be disassembled for transportation and storage. The size and weight of even a moderately sized Newtonian can prove taxing for a mount, particularly a GoTo mount.

Collimation (precise alignment of the optical path) of a Newtonian reflector is a challenge for a beginner and must be performed regularly to insure quality performance. Having access to an experienced friend or local astronomy club is almost certainly required for the new owner of a Newtonian reflector.

Newtonian reflectors excel at deep sky viewing in larger apertures, but in my opinion the longer tube lengths required for larger apertures, and the added weight this entails, do not make them well suited to most GoTo mounts.

Schmidt–Cassegrain Catadioptric   (Figure 2.15) Schmidt–Cassegrain

telescopes, or SCTs as they are commonly known, combine the best elements of refractor

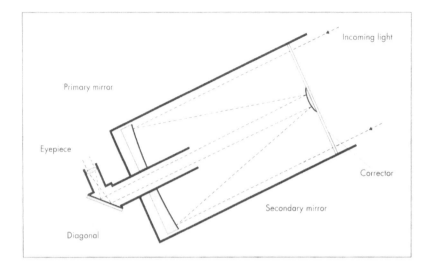

Figure 2.15. The Schmidt–Cassegrain telescope design.

and reflector designs to produce the most compact optical tubes available for any given aperture. Catadioptric or compound scopes possess a combination of lenses and mirrors. The SCT design is the most popular line of serious amateur telescopes sold today. Celestron perfected the process of mass-producing the Schmidt–Cassegrain design in the late 1960s, and popularized the design through the 1970s.

At the front of the tube is a lens known as the corrector plate. The corrector plate allows for improved performance from a spherical primary mirror, reducing the cost of producing a high-quality instrument. Mounted in the center of the corrector plate is the secondary mirror that directs the light to the back of the tube, through a hole in the primary mirror, towards the eyepiece. Like refractors, SCTs use diagonals to provide convenient positioning of the eyepiece. Due to the compact size of the SCT design, the placement of the eyepiece at the rear of the scope provides comfortable viewing for most orientations of the optical tube. Additionally, the corrector plate protects the mirrors from corrosion, greatly increasing the longevity of their coatings.

The folded light path provided by the corrector plate and mirrors allows the focal length of the scope to be about five times the length of the optical tube. In refractors and most Newtonians the optical tube is about the same length as the focal length. As a comparison, the focal length of the NexStar 8 (an SCT design) is about 80 inches (200 cm), while its optical tube is about 16 inches (40 cm) long. A refractor or Newtonian with a similar focal length would sport a tube nearly 7 feet (2 meters) in length!

There are some tradeoffs, though. SCTs are more expensive than Newtonian reflectors of similar aperture, although lower in cost than comparable refractors. They do require precise collimation, but fortunately the process is much less complicated than with a Newtonian. The closed-tube design, resulting from the corrector plate, increases cooldown time, particularly in larger models. Some models also exhibit excessive image shift when focusing. That is to say that focusing will actually cause the object in the field of view to move up, down, left, or right. This is due to the method that focus is achieved with in most SCTs – the primary mirror slides up and down to move the focal plane. While other manufacturer's scopes exhibit a good deal of mirror shift, Celestron's focusing mechanism minimizes this dramatically.

The SCT design is a great all-around performer and is well suited to use on a GoTo telescope mount. The compact and robust design results in a fine instrument that can provide exhilarating views for many years to come.

Maksutov–Cassegrain Catadioptric (Figure 2.16) Similar in physical design to SCTs, Maksutov–Cassegrain telescopes, or Maks as they are commonly known, use a lens and mirrors to focus gathered light. The main difference you will note is the shape of the lens, typically called the corrector or meniscus lens. In a Mak, the meniscus is a deep, curved shape rather than the flat corrector plate found in an SCT. In most Maks the secondary mirror is a reflective coating applied directly to the back of the meniscus.

Generally Maks are designed with a long focal length and slow focal ratio of f/12 or higher. This makes them well suited to viewing the planets and fine details on the Moon. The resulting narrow field of view does exhibit some limitations when viewing deep sky objects, particularly extended open clusters and galaxies. Also, their slow focal ratio results in very long exposure times for photographs of deep sky objects. In addition to their long focal length, the relatively small size of the secondary mirror (spot) provides slightly higher-contrast views of the Moon and planets than those obtainable with SCTs.

Maks are generally more expensive than SCTs of similar aperture, in price falling somewhere between SCTs and refractors. Also, larger Maks are rare, most models being 5-inch or smaller. The meniscus is thicker than the corrector plate on SCTs, so cool-down time is increased.

Maks are generally very low-maintenance. Most models do not provide for user collimation and are expected to keep their alignment indefinitely. Like all reflector-type scopes, they exhibit no noticeable chromatic aberration. These factors, coupled with excellent

Figure 2.16. The Maksutov–Cassegrain telescope design.

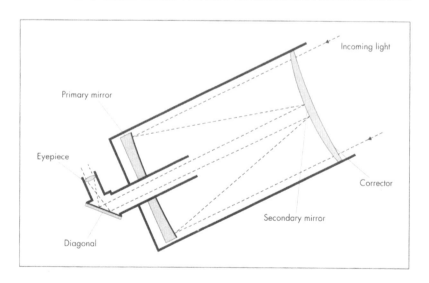

planetary views and very compact design, make the Maksutov–Cassegrain a good choice for someone looking for the advantages of a fine refractor, in a less expensive, more convenient package.

Finder Scopes

Finder Scopes (Figure 2.17) The narrow field of view of a telescope makes it very difficult to aim accurately at a desired location in the sky. To overcome this problem, a finder scope, or finder, is mounted on the side of the main telescope. When correctly aligned, the finder scope and the main scope both point at the same place in the sky. The lower magnification and wider field of view of the finder make it easier to locate and point the scope at any given position in the sky. When you look in the eyepiece of the main scope your target should be visible.

There are two basic types of finders. Traditional finders are small refractor scopes with crosshairs in the eyepiece. The alternatives are "zero-power" or "unity" finders. Unity finders project a red dot or bull's-eye onto a small window. By looking through the window, you can easily aim the scope with the dot or bull's-eye.

The common refractor-style finder is designated in the same way as binoculars. A 6×30 finder has an aperture of 30 mm and magnifies 6 times. Larger aperture makes it easier to find fainter objects, although 50 mm is a practical limit for most uses. There are two basic types of refractor finders. The most common is a straight-through finder where the eyepiece points straight back from the tube of the finder. Most straight-through finders produce an image that is upside-down. Moving the telescope in the right direction to aim with the finder can be a little confusing at first, but after a little while it becomes second nature. To match the view in the finder to your star charts, you simply turn the chart upside-down. Alternatively there are correct-image straight-through finders that simplify matters further. While providing a more natural right-side-up view, straight-through finders are slightly disadvantaged by having reduced light throughput (dimmer images) and added expense.

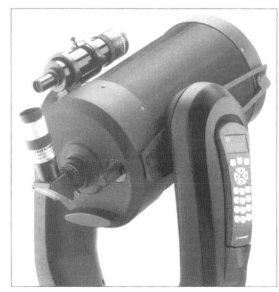

Figure 2.17. NexStar 8 GPS with 9×50 finder scope. Photo courtesy Celestron.

The second type of refractor-style finder uses a diagonal and is known as a right-angle finder. While a right-angle finder produces more comfortable positioning of the eyepiece, most models produce a left-to-right mirror image that is difficult to match to printed star charts. A few more expensive models do provide a correct image and represent the height of traditional finder scope design.

Unity finders are particularly easy to use since they simply project a target on your naked-eye view of the sky. Aiming the scope could not be easier. They do present some disadvantages. First, since they do not collect light or magnify the view, you are limited to stars you can see with the naked eye. Second, the lack of magnification can make it difficult to aim precisely at a star and have it show in a narrow-field, high-magnification view in the main scope. But unity finders are well suited to use with GoTo telescopes, since typically we only point manually at two bright stars during the initial setup and the computerized hand control takes it from there.

With any type of finder aligning them with the telescope is a simple task best performed during the day. Looking through the main scope with a moderately powered eyepiece, aim your telescope at a distant object such as a ceramic insulator on an electric pole. Then simply turn the adjustment screws on the finder scope until it is also aimed precisely at the same point. Check to be certain that the finder is well secured and you are finished.

Telescope Mounts

The mount is generally comprised of two sections – the part that holds the telescope and allows it to be pointed in any direction, and the tripod. The section that holds the scope and moves is referred to as the mount, while the tripod is considered separately. In some cases there is no tripod, or the tripod is optional, but in many cases the mount and tripod are sold as a single unit.

It is absolutely critical that the mount provides excellent stability for the telescope in use. An unstable or undersized mount will make the view in the eyepiece shake terribly at the slightest touch or in the lightest wind. When trying out a telescope, perform the "rap test": using a high-powered eyepiece, focus on a distant object and tap the eyepiece end of the scope. All vibration should dampen within no more than 2 or 3 seconds; less would be ideal.

Also check that the movement of the mount is smooth and responsive. As the Earth is turning, all objects move across the sky. The mount must be continually adjusted to keep an object in the field of view. A mount with jerky movement will prove difficult, if not impossible, to use when tracking the sky.

Finally, consider complexity and portability. Some mounts require just a few minutes to set up, while others can require a half-hour or more. Collapsible tripods are common and provide good portability, while large Dobsonian mounts may require a van or truck to transport. If you want a scope that can be set up on a whim and is ready for service in minutes, choose accordingly. If ultimate stability or an extremely large optical tube requires a more heavy-duty mount, portability might take a back seat.

Mounts can be categorized into two basic types: altitude–azimuth (alt–az) and equatorial.

Altitude–Azimuth Mounts (Figure 2.18) The alt–az mount allows the optical tube to be pointed up/down and left/right. This arrangement is very simple to use and well suited for beginners, but it generally has one major drawback – objects in the sky don't move up/down or left/right. As discussed earlier, they follow an arc from east to west as the Earth turns on its axis at sidereal rate. To track an object with an alt–az mount, we

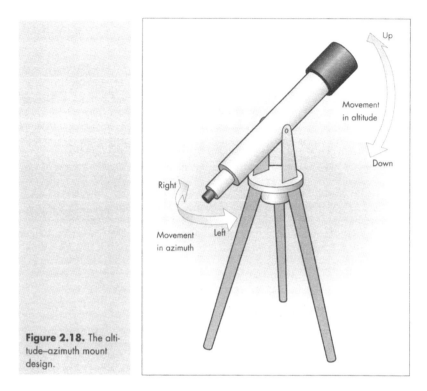

Figure 2.18. The altitude–azimuth mount design.

must periodically move the altitude and azimuth axis to keep the object in the field of view. If the field of view in the eyepiece is several degrees, we will be adjusting every couple of minutes. If we are examining the Moon or a planet at high magnification, a small field of view, we will be adjusting every couple of seconds.

Generally, an alt–az mount cannot be motor-driven with much success, unless a computer controls the motors to provide the correct movement in both axes to allow tracking the sky at sidereal rate. We find such computerized control in today's GoTo telescopes.

Dobsonian mounts, or Dobs as they are commonly called, are a variation on the alt–az mount. Created by John Dobson in the 1970s, they are the most popular mounts for moderate-to-large-sized Newtonian reflectors. The low center of gravity of a Dob makes for a very stable platform for even the largest scope. They are rarely motorized, but their smooth motion allows for easy tracking by simply nudging the end of the scope.

Equatorial Mounts (Figure 2.19) An equatorial mount takes a different approach to providing motion for the optical tube. One of the axes is aligned with the axis of the Earth and can be turned opposite the motion of the Earth to keep an object in the field of view. This axis is known as the right ascension (RA) axis since it allows us to point at any right ascension in the sky. The second axis is known as the declination (Dec) axis and provides the ability to point towards any declination above or below the celestial equator.

Tracking the sky at sidereal rate is relatively easy with an equatorial mount; we simply adjust the RA axis periodically. Many equatorial mounts can be equipped with a motor to

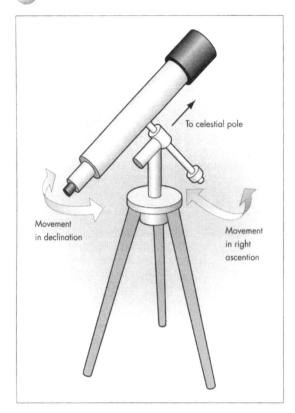

To celestial pole

Movement
in declination

Movement
in right
ascention

Figure 2.19. The German equatorial mount design.

drive the RA axis at sidereal rate, providing automatic tracking of any object in the sky. In this fashion, visual observations become much more convenient and long-exposure astrophotography becomes possible.

This convenience does come at the price of added complication. While we can take an alt–az scope out, extend the legs, and start observing, with an equatorial mount we need to align the RA axis with the Earth's axis. For observers in the Northern Hemisphere, we simply point the RA axis at Polaris, the North Star. Some equatorial mounts even include a small telescope in the center of the RA axis to make short work of this. Observers in the Southern Hemisphere have a much harder time here, as there are no bright stars near the south celestial pole. A rough alignment using just the sighting method is sufficient for visual work, but astrophotography requires much more exact alignment to prevent drifting that will turn pinpoint stars into little dashes in the final exposure.

Fork-mounted alt–az scopes can be equatorially mounted for astrophotography. Using a wedge, the forks (Figure 2.20) are tilted so that the azimuth axis is pointed at the celestial pole, thus becoming the RA axis. A wedge is commonly used with fork-mounted SCT scopes to allow for long-exposure astrophotographs.

Piers A pier is not a complete mount, but rather takes the place of the tripod. Generally we then attach an equatorial mount on top of the pier. Most piers are permanent fixtures, either in the owner's backyard or in their home observatory. By permanently mounting

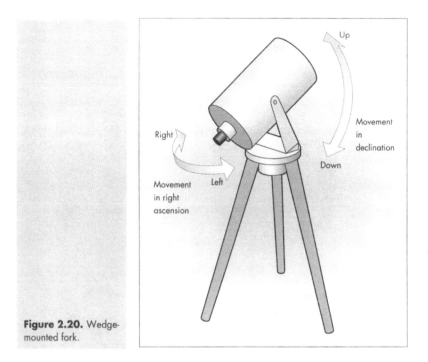

Figure 2.20. Wedge-mounted fork.

the pier in the ground, we have an extremely stable mount that provides very accurate alignment of the RA axis. The scope can be removed and stored indoors when not in use.

Some piers are portable, providing added stability for large telescopes when used in the field. In most cases they are lighter than similarly stable tripods.

The Eyepiece End of Things

While initially the telescope gets most of the attention, before long you will be out shopping for additional eyepieces. We will cover eyepiece design and selection in more detail in Chapter 8, but here we will discuss what the eyepiece does and its key characteristics.

The purpose of the main telescope is to gather light and bring it to a focal point, hopefully without altering the image in any way. It is then the job of the eyepiece to magnify the image to a suitable scale. For this purpose, eyepieces are available in a variety of focal lengths, usually displayed on the outside of the barrel. As discussed earlier, we calculate the magnification an eyepiece gives by dividing the focal length of the main scope by the focal length of the eyepiece. For a telescope with a focal length of 1000 mm, a 20mm eyepiece yields 50×, while a 10mm eyepiece yields 100×. Thus, shorter-focal-length eyepieces provide higher magnifications.

Besides focal length, eyepieces have other vital statistics. Eye relief, or the distance your eye may be from the eyepiece and still take in the entire view, is one. With most eyepiece designs, the shorter the focal length, the shorter the eye relief. Some short-focal-length eyepieces require you position your eye uncomfortably close to see the object in the field

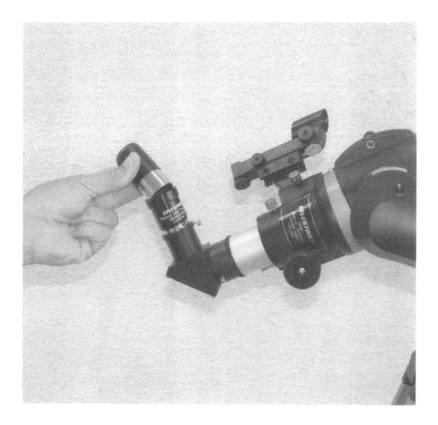

Figure 2.21. Inserting an eyepiece into a Barlow mounted in a diagonal.

of view. If you must wear or prefer to wear eyeglasses when viewing, you would do well to choose eyepieces with longer eye relief.

Another characteristic of eyepieces is apparent field of view or AFOV. AFOV of less than about 50° results in a view that resembles looking through a straw. Wider AFOV provides a panoramic view that is preferred by most observers. The actual width of sky seen through a given eyepiece–telescope combination is referred to as the true field of view or TFOV. The approximate TFOV can be calculated by dividing the AFOV of the eyepiece by the magnification given with that eyepiece. Taking our 1000mm telescope and 10mm eyepiece, we have a magnification of 100×. If the 10mm eyepiece has an AFOV of 50°, then the TFOV will be 50 divided by 100, or $\frac{1}{2}°$.

A common accessory known as a Barlow lens is used to increase the magnification given by any eyepiece. The typical Barlow doubles the magnification when used directly in front of the eyepiece and triples the magnification if used between the telescope and the diagonal. When used as shown in Figure 2.21, a 10mm eyepiece doubles the magnification; in a 1000mm telescope this results in 200×. In addition to effectively doubling your eyepiece collection, higher magnifications are achieved without resorting to short-focal-length eyepieces with their typically short eye relief.

Conclusion

What you have read in this chapter may have been the first step on your journey of learning about the cosmos and the equipment that will let you observe it. There are many excellent books that can take you further, but I would recommend a few in particular. For learning more about general astronomy, take a look at *Nightwatch* by Terence Dickinson or *The Night Sky* by David Chandler. *Exploring the Night Sky* by Terence Dickinson is a wonderful introduction to astronomy for younger kids. For an excellent overview of equipment and accessories, I recommend *Star Ware* by Philip Harrington or *Choosing and Using a Schmidt Cassegrain Telescope* by Rod Mollise. Additionally, a subscription to either *Sky and Telescope* or *Astronomy* magazine will let you keep up with the newest developments in astronomy and equipment.

Part of the enjoyment of viewing astronomical objects is understanding a little about the objects we observe. Astronomy is part intellectual and part visual. Knowing that the light that you are seeing now traveled for tens of millions of years to reach you adds to the excitement of detecting that faint smudge in the eyepiece. The humbling realization that we live on such a tiny little speck in just one corner of the universe seems to make our daily troubles fade in comparison. And yet, insignificant as we may seem, just the fact that we can begin to understand our place in the cosmos is truly unique. Welcome to a special crowd.

Chapter 3
Overview of the NexStar Line

All NexStar telescopes share the same basic operation. Once you learn to use one model of NexStar, you can use any model of NexStar – from the little NexStar 60 GT to the sophisticated NexStar 11 GPS. There are various options for the initial alignment of the telescope, but they all rely on you to center two stars that we refer to as "alignment stars". After you have centered the two alignment stars the telescope now has a working model of the sky. You may then select objects and the telescope will slew to them at the press of a button.

The simple operation of the NexStar system is one of its great strengths. To slew to most objects, you press a catalog button, type in the number of the object, and then press the Enter button. For example, to GoTo the Andromeda galaxy (M31), you would press the M button (Messier), type in "031", and then press the Enter button. Simplicity at its finest.

And that simplicity pays off for a new owner. I ordered my NexStar 80 GT in October 2000 and it experienced first light on 6 November after less than an hour for assembly and familiarity. As described in the Introduction, it was a good night that exceeded my expectations. Even today, I feel the NexStar 80 GT is one of the absolute best beginner scopes on the market, as well as the best value in the low-priced GoTo market.

Another characteristic shared by the NexStar line is stability. Telescopes in the under-$400 range are typically supplied with tripods that shake and dance at the slightest touch and take many seconds to come to a rest. And the typical alt–azimuth or equatorial mounts on such entry-level scopes are notoriously difficult to aim precisely at any point in the sky. Such is not the case with the entry-level NexStar models. Motion of the telescope with the directional buttons is smooth and there is very little play in the mechanics of the mounts. The supplied or recommended tripods are well matched for the weight of the various models. The advanced NexStar models are also very stable, and motion control and tracking are even more precise. All models deaden vibration in less then 2 or 3 seconds, most models almost immediately.

No one telescope can meet everyone's requirements nor the various desires of the more dedicated amateur astronomer. It is not uncommon for the more afflicted among us to own two or more telescopes! To meet various needs, Celestron has built a wonderfully diverse line of NexStar telescopes, each with different strengths and weaknesses.

Entry-level NexStars – the "Little NexStars"

The NexStar 60, 80, 114, and 4 comprise the models sometimes called the little NexStars. They are all priced for those looking for their first scope or for those looking for a second, smaller, more portable scope to complement a larger, more expensive instrument. They all share a similar mechanical design and the same computerized hand control. Their GoTo accuracy is very good, but tracking is really designed for visual use only. In other words, objects do not stay centered in the field of view for extended periods, and thus these scopes are limited in their astrophotography capabilities.

Despite this weakness, they are wonderfully capable of opening the night sky and showing its many treasures. If you think you might be interested in astronomy, but you haven't yet made the jump to a telescope, any of these models would be a good way to start.

NexStar 60

The NexStar 60 was initially sold in two configurations, the ST and GT models. Both models include an adjustable tripod. The ST model had no motors and no hand control. Celestron intended to sell the motors and computerized hand control as an owner-installed option, but they decided that the mechanical alignment of the motors and gears required precision that most owners would not achieve and they never released the upgrade kit. If you were unfamiliar with this limitation and purchased the ST model, I suggest you contact Celestron and see if they will upgrade your telescope at their factory.

The GT model (Figure 3.1) is the only NexStar 60 currently being sold. It features the optical tube of the classic beginner's scope, a 60mm achromatic refractor. The motor drives and computerized hand control provide full GoTo capability. After it slews (moves) to an object, it tracks it automatically, following the movement of the sky due to the rotation of the Earth.

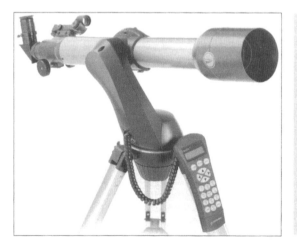

Figure 3.1. NexStar 60 GT. Photo courtesy Celestron.

NexStar 60 GT at a Glance

Optical system: achromatic refractor
Approximate street price: $250 (includes tripod)
Aperture: 60 mm
Focal length: 700 mm
Focal ratio: f/12
Supplied eyepieces: 4 mm (175×),
10mm (70×), 20mm (35×): All are MA design
Maximum magnification: 120×
Maximum field of view: 2.5°
Magnitude limit: 10.5
Resolution limit: 2.3 arc second
Finder: 1× power red-dot
Objects in HC database: 4033
Weight (includes tripod): 9.5 lb (4.3 kg)

As with all NexStars except the 8/11 GPS, power is supplied by 8 AA batteries. In the case of the NexStar 60, the batteries are held in a small pouch that hangs from the tripod. AA batteries don't last long and are an expensive way to operate the scope; see Chapter 8 for power supply suggestions.

Due to the length of the tube it is not always possible to point at objects high in the sky, near to the zenith. The telescope might make contact with a tripod leg before it reaches its target. For that reason, the GT hand control has set as its default a maximum slew limit of 65° altitude. In other words, it warns when asked to GoTo an object at an angle above 65° above the horizon. You can change the slew limits, but if the telescope does contact the tripod, it looses track of where it is pointed in the sky and you must perform the initial alignment again. It is possible to override the slew limits and GoTo an object when the scope will land between two of the tripod legs; see Chapter 5 for more details.

An aperture of 60 mm does have severe limits regarding what can be seen. On nights with good transparency and seeing you can expect wonderful views of the Moon,

Old Versus New GT Hand Control

The first thing you should do when you get any of the little NexStars (the 60, 80, 114, and 4 models) is to insure that you have the second version of the GT computerized hand control. This can easily be confirmed right after applying power – if you have Auto, Two-Star, and Quick alignment options, it is the new hand control. Celestron released this updated GT hand control in December 2001. Little NexStars manufactured prior to this have several problems. Below is a list of the main complaints and their status in the new hand control:

- No official RS-232 support due to several problems such as hand control lockup and inability to report coordinates above 12 hours Right Ascension or negative Declination – the new hand control includes the NexStar GPS RS-232 protocol, exhibits no lockup problems, and accurately reports the current coordinates of the telescope. Use of the GPS control protocol means that any PC/Mac/Palm/PocketPC program that works with the NexStar GPS models works with this new hand control. Many astronomy programs do not support the old GT hand control.
- Won't save observation locations in the Eastern Hemisphere – fixed.
- Won't save backlash settings – fixed .
- Cordwrap feature not functional – still not included, but at least it has been removed from the menus to prevent confusion about its lack of operation.
- Occasional lockups and runaway slews – fixed.
- Very touchy keypad that often leads to double key entries – fixed.
- Light Control feature not functional – fixed .
- Tracking Rate defaults to Solar – fixed, the default is now Sidereal.
- Error in the coordinates of M2, M10, and M110 – fixed.

Old Versus New GT Hand Control *(continued)*

Several additional features were added to the new hand control as well:

- Two-Star and Quick Align alignment modes – the original hand control only has Auto Align.
- All alignment methods have been improved as the scope now permanently stores setup information such as observing location and time zone.
- User-definable slew limits have been added to set the maximum and minimum altitude (angle above the horizon) that the scope should be allowed to move to. You are able to override the limits you set, but they prove to be valuable in preventing the telescope from making contact with the tripod.
- User-definable filter limits have been added to limit the object catalogs to only those objects within your minimum and maximum altitude boundaries.
- Faster slewing to objects during a GoTo. When a NexStar telescope slews to an object, it first moves at the fastest rate until it gets near the object. Then it slows to a crawl for the final approach. With the new GT hand control the fast slew gets closer to the object and the slower final approach is shorter.
- Manual slewing with the arrow keys is more responsive as the new hand control applies backlash compensation more smoothly.

The upgraded hand control is quite an improvement on an already good product. If you receive a NexStar telescope with the old version (the easiest test is to look for the Two Star alignment option right after powering up), contact Celestron via Technical Support at their web site (www.celestron.com) to see about a replacement.

Jupiter's two main bands and four moons, and Saturn's rings will display as a single distinct ring. Note that the highest useable magnification is about 120×, which leads to very tiny images of even these large planets. You will rarely see any features on Mars through the N60. Venus will show its phases, similar to our view of the Moon. Colorful double stars are a treat in the N60, as are bright open clusters. Most nebulae and galaxies are beyond the reach of this scope, although the brighter ones can be easily detected at a dark site away from all light pollution.

The supplied eyepieces are this telescope's greatest weakness; refer to Chapter 8 for recommendations on replacements. The 4mm is nearly useless; it provides much too high magnification that will be just barely useful when viewing the Moon. The 10mm is OK to get you started at a reasonable magnification for your first views of the Moon and planets as well as many double stars. The 20mm provides a nice wide field for open clusters and your first attempts at those brighter galaxies and nebulae.

Perhaps the best reason for purchasing a NexStar 60 GT is the low price coupled with the potential for replacing the 60mm optical tube assembly (OTA) with a larger, higher-quality OTA. The mount on the NexStar 60 GT is identical to the NexStar 80 GT and both can make use of the Baader Bracket (see Chapter 13) to provide GoTo performance for any small OTA.

In all, you get a lot of functionality for relatively little money when you buy a NexStar 60 GT, but I don't really recommend it for most. If you save just a little longer you can afford the NexStar 80 GT, which I consider to be a much better buy.

NexStar 80

The NexStar 80 was also originally offered in two configurations, the HC and GT models. The HC model was fully motorized, but the electronic hand control was fully manual. This

Figure 3.2. NexStar 80 GT. Photo courtesy Celestron.

allows you to point the scope anywhere desired and manually track objects using the four directional arrow buttons. The HC model can be updated by purchasing the optional computerized GT hand control, Celestron part number 93961. At the time of this writing it is available for about $80 at Astronomics (http://www.astronomics.com) and other vendors stock it as well. Due to potential difficulties locating a GT hand control, and the added expense of buying things separately, it is best to avoid the HC model should you happen to find one available.

As it turns out, most of the NexStar 80s sold were either the GT model (Figure 3.2), with a computerized hand control, or the HC model bundled with the GT hand control option. Both models include an adjustable tripod. As mentioned in the Chapter 1, the NexStar 80 GT was my first NexStar telescope and I certainly have a sweet spot for this wonderful little scope.

Power is supplied by 8 AA batteries in a small pouch that hangs from the tripod. See Chapter 8 for power supply suggestions that won't tax your pocketbook, as you will go through AA batteries at an alarming rate.

As with the NexStar 60, the length of the tube can interfere with the ability to point at objects high in the sky near the zenith. Refer to Chapter 5 for techniques in dealing with this situation by utilizing Slew Limits.

The optical tube is the very popular short-tube 80mm achromatic refractor imported from China under many different

NexStar 80 GT at a Glance

Optical system: achromatic refractor
Approximate street price: $350 (includes tripod)
Aperture: 80 mm
Focal length: 400 mm
Focal ratio: f/5
Supplied eyepieces: 10 mm SMA (40×), 25 mm SMA (16×)
Maximum magnification: 160×
Maximum field of view: 4.4°
Magnitude limit: 11.3
Resolution limit: 1.8 arc second
Finder: 1× power red-dot
Objects in HC database: 4033
Weight (includes tripod): 11 lb (5 kg)

brand names. The 80mm aperture gathers almost 80% more light than the NexStar 60. This brings more faint objects into view. Many more galaxies and nebulae are visible in the NexStar 80 than the 60. The added aperture also provides more detail on objects. You will be able to see smaller craters and other features on the Moon. On very clear and steady nights you will see substantially more detail on Jupiter, although it will still be mostly just a cream-colored disk with two or three dark bands. On the very best nights from a dark location you

Tasco StarGuide Telescopes

In 1998, Tasco purchased Celestron and it is presumed they provided most of the funding for the development of the NexStar line. A series of models from Tasco known as StarGuide were released about the same time as the little NexStars. They are virtual clones of the little NexStar scopes except for two things.

First, the hand control has only about 1800 objects as compared to the more than 4000 in the little NexStars. This is not a terribly important issue, since you can still go directly to any object by entering its right ascension and declination.

Second, the hand control is the same as the original GT hand control and exhibits all the problems described in the sidebar Old versus New GT Hand Control. Tasco did not upgrade their hand control, but the Tasco telescopes are compatible with the NexStar GT hand controls. You can upgrade your StarGuide scope by purchasing the NexStar GT hand control separately, although they are getting hard to find.

Tasco has since gone out of business and Celestron is now a privately owned company once more, but Celestron did not inherit responsibility for the StarGuide line.

can make out the Cassini division on the extremities of Saturn's rings. Prior to the dust storm that covered Mars in 2001, I was easily able to make out dark features on the surface.

The supplied eyepieces will get you started, but again they are the weakest components of the scope. With the short 400mm focal length of the NexStar 80, the 10mm eyepiece results in a magnification of only 40×. At a minimum you will want a Barlow lens to increase the magnification to 80× when using the 10mm eyepiece. Eventually you will definitely want a shorter focal length eyepiece to increase the magnification at or near the theoretical maximum of 160×. I find that the scope is fully capable of supporting this magnification. This provides nice views of the Moon and planets. Refer to Chapter 8 for recommendations on additional eyepieces.

The greatest strength of the NexStar 80 is the wide-field views it delivers with its short focal length. Using the supplied 25mm eyepiece results a in true field of view of more than 3°. This allows you to take in the entire Andromeda Galaxy (M31), the Pleiades (M45), all of the Great Orion Nebula (M42), and other of the most spectacular deep sky objects. Touring around the many treasures in the constellation Sagittarius – the center of our galaxy – is truly a delight. With a 40mm eyepiece you can view nearly 4.5°, rivaling the field of view in many binoculars.

But that same short focal length and the NexStar 80's achromatic design lead to false color on bright objects such as the Moon, Jupiter, and Venus. It also makes it necessary to use a short, 5mm eyepiece **and** a Barlow lens in order to reach its theoretical magnification limit of 160×. Nonetheless, it is a very good performer, quite portable and my best recommendation for beginners on a tight budget.

NexStar 4

The NexStar 4 with its Maksutov-Cassegrain optical tube is the most portable of all the NexStar line. It includes the GT computerized hand control stowed in its single fork arm (Figure 3.3). The basic model does not include a tripod, but can be used without when set

Figure 3.3. NexStar 4 GT. Photo courtesy Celestron.

on any flat surface. The NexStar 4 is occasionally sold in a bundle with a "wedge-pod" – a tripod with an equatorial wedge built into the head – which I would recommend most buyers to seriously consider. Power is supplied by 8 AA batteries held in the base of the telescope or any external 12-volt power source.

While all other compound telescopes sold by Celestron – the Schmidt–Cassegrains – are made in their factory in California, the NexStar 4, like the other little NexStars, is manufactured in China. Quality of these scopes has been quite good, although some of the earliest NexStar 4s to reach owners had a problem with a misplaced secondary baffle; refer to Chapter 14 for details.

The NexStar 4 is a great performer. Its 4 inch aperture allows it to pull in $2\frac{1}{2}$ times more light than the NexStar 60, allowing it to see a full magnitude deeper into space. Its long focal length means higher magnifications with longer, more comfortable eyepieces. It is a great little scope for the Moon and planets. The amount of detail seen on the Moon can occupy your

NexStar 4 GT at a Glance

Optical system: Maksutov–Cassegrain
Approximate street price: $500 (no tripod)
Aperture: 4 inches (102 mm)
Focal length: 1325 mm
Focal ratio: f/13
Supplied eyepiece: 25 mm SMA (53×)
Maximum magnification: 200×
Maximum field of view: 1.3°
Magnitude limit: 11.7
Resolution limit: 1.4 arc second
Finder: 1× power red-dot
Objects in HC database: 4033
Weight (no tripod): 11 lb (5 kg)

viewing for several nights each month. On steady nights Jupiter shows slight detail in its bands – occasionally you will detect the Great Red Spot – and the Cassini Division in Saturn's rings is an easy target under dark skies. On favorable passes, studies of Mars will reveal subtle details not seen in the smaller NexStar 80. Deep sky objects also provide good views in the NexStar 4, especially compact open clusters, planetary nebulae, and the brighter globular clusters.

The longer focal length and high focal ratio do have some drawbacks. The maximum field of view with the NexStar 4 is just 1.3°, or about $2\frac{1}{2}$ times the size of the full Moon. And the f/13 focal ratio means long exposures if you try any prime focus astrophotography. But, for visual use, the NexStar 4 is a true joy, even with its limitations.

The NexStar 4 sports one feature specifically designed for astrophotography, a port on the back of the scope for attaching a camera at prime focus. This is incorporated with a built-in "flip mirror" diagonal that allows you to more easily center and focus an object in the eyepiece, then flip the mirror to pass the image through to the camera. See Chapter 11 for more details on astrophotography.

NexStar 114

The NexStar 114 is the largest aperture of Celestron's entry-level GoTo scopes. In fact, considering that the central obstruction (secondary mirror) is proportionally smaller than the SCT design, the NexStar 114 gathers almost as much light as the NexStar 5 and at just one-third the cost. So why doesn't the 114 eliminate the market for the NexStar 5? The answer is multifaceted, but lies largely in the much higher optical quality and the greater portability of the NexStar 5.

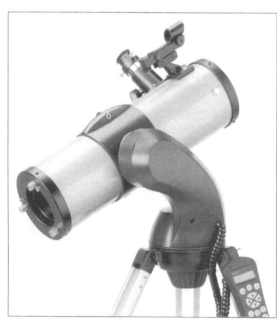

Figure 3.4. NexStar 114 GT. Photo courtesy Celestron.

Like its smaller sibling, the NexStar 80, the NexStar 114 was originally sold in an HC and a GT model. The HC model, motorized with manual hand control, could be upgraded by purchasing an optional computerized GT hand control, but these are increasingly hard to find. Both models include an adjustable tripod. Currently you should only find the NexStar 114

NexStar 114 GT at a Glance

Optical system: Newtonian reflector
Approximate street price: $440 (includes tripod)
Aperture: $4\frac{1}{2}$ inches (114 mm)
Focal length: 1000 mm
Focal ratio: f/8.8
Supplied eyepiece: 10 mm SMA (100×), 25 mm SMA (40x)
Maximum magnification: 225×
Maximum field of view: 1.75°
Magnitude limit: 12
Resolution limit: 1.2 arc second
Finder: 1× power red-dot
Objects in HC database: 4033
Weight (includes tripod): 15.5lb (7kg)

GT (Figure 3.4) available. A pouch with 8 AA batteries hangs from the tripod to supply the required power, or a better alternative is an external 12-volt power source such as those suggested in Chapter 8. And like the NexStars 60 and 80, the tube of the 114 can also make contact with a tripod leg. Refer to Chapter 5 for tips to deal with this limitation by using Slew Limits.

The NexStar 114 is a short-tube Newtonian reflector with a spherical mirror. The mirror's actual focal length is quite short, but a corrector lens is used at the eyepiece port to lengthen it to 1000 mm. While providing a reasonably long focal length, unfortunately that lens can also lead to much frustration for new telescope owners. It seriously complicates the process of collimation. Collimation is the process of precisely aligning (collimating) all of the optical components to insure the focal plane of the telescope arrives perfectly aligned with the focal plane of the eyepiece. We will discuss collimation in Chapter 9.

The $4\frac{1}{2}$-inch aperture brings many deep sky objects (DSOs) into view. Hundreds of galaxies, star clusters, and nebulae are within reach of the NexStar 114. The brighter, show-case DSOs, such as the Ring (M57) and Dumbbell (M27) nebulae, the Great Orion Nebula (M42), and the Andromeda Galaxy (M31), will display wonderful amounts of detail under dark, clear skies. You will be able to resolve some stars on the brighter globular clusters like M13 and M22. The NexStar 114 is a great beginner's scope for DSOs.

With a well-collimated NexStar 114, expect some very exciting planetary and lunar views through the eyepiece. Endless craters, rills, and mountain ranges await you on the Moon. Nights of steady seeing and 200× magnification will bring out some details in the two prominent bands on Jupiter, occasional views of the Great Red Spot, and also a clear look at a third band. Similar conditions will show Saturn as a golden color with a clearly defined Cassini Division in the rings.

The eyepieces supplied are budget SMA models that will get you started, but you will gain much by upgrading. Additionally, a Barlow lens will be of good use with the NexStar 114. See Chapter 8 for eyepiece recommendations.

Advanced NexStars – the SCT Models

The remaining NexStar models are priced considerably higher and appeal to more com-mitted amateur astronomers, or at least those willing to consider a higher budget. All of the remaining models feature Celestron's excellent Schmidt–Cassegrain optical design and are made in the Celestron factory in California. Schmidt–Cassegrain telescopes (SCTs) are wonderfully versatile and compact for any given aperture. All of Celestron's SCTs include hand-figured optics – the secondary mirror is ground to its final shape by hand, allowing very high optical quality in the final product. Additionally, the mirrors are surfaced with Celestron's Starbright coatings – a special five-step multi-layer process that results in extremely high reflectivity. The corrector plate – the lens on the front – sports high-quality anti-reflective multi-coatings on the front and back surfaces. The optics in the NexStar SCTs have been universally praised in review after review.

NexStar 5 and 8

As discussed in the first chapter, these are the granddaddies of all NexStars. Recently replaced by the NexStar 5i/8i models, the NexStar 5 and 8 are a very good buy as they are

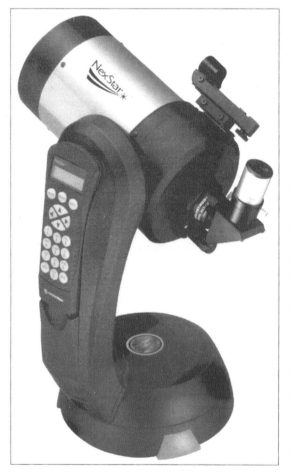

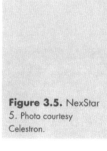

Figure 3.5. NexStar 5. Photo courtesy Celestron.

available at closeout prices or on the used market. The NexStar 5 (Figure 3.5) is sometimes offered with a tripod, sometimes without. The NexStar 8 (Figure 3.6) is generally offered with a tripod. Both include a computerized hand control with an 18 000-object database.

NexStar 5 at a Glance

Optical system: Schmidt–Cassegrain catadioptric
Approximate street price when new: $1050 (includes tripod)
Aperture: 5 inches (127 mm)
Focal length: 1250 mm
Focal ratio: f/10
Supplied eyepiece: 25 mm Plössl (50×)
Maximum magnification: 250×
Maximum field of view: 1.4°
Magnitude limit: 12.3
Resolution limit: 1.1 arc second
Finder: 1× power red-dot
Objects in HC database: 18 000
Weight (no tripod): 17.6 lb (8 kg)

The NexStar 5 and 8 were designed primarily as visual use scopes and were not intended for astrophotography. Limited astrophotography is possible, but manual guiding is required and the mediocre tracking accuracy of the NexStar 5 and 8 can make this very tedious. The new NexStar 5i and 8i were improved to overcome this limitation.

The NexStar 5 and 8, as mentioned in the introduction, both use the same one-arm fork mount. It is a very stable platform and makes for the lightest, most portable 5- and 8-inch GoTo scopes on the market. These are truly "grab and go" telescopes with ample aperture, great optics, and accurate GoTo capability. If what you are looking for is the lowest price for stunning views in a scope that you can set up in minutes, you can't go wrong with either of these scopes.

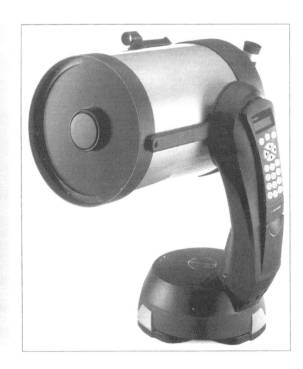

Figure 3.6. NexStar 8. Photo courtesy Celestron.

Power is supplied either by 8 AA batteries that store in the top of the base, or any 12-volt power source. Both scopes can be used set on a flat, stable surface or mounted on a tripod. I recommend the tripod for added stability and security, but without a tripod the NexStar 5 fits in a backpack and can accompany you on treks into the wilderness. The eyepiece supplied is of good

NexStar 8 at a Glance

Optical system: Schmidt–Cassegrain catadioptric
Approximate street price when new: $1300 (includes tripod)
Aperture: 8 inches (203 mm)
Focal length: 2032 mm
Focal ratio: f/10
Supplied eyepiece: 40 mm Plössl (51×)
Maximum magnification: 400×
Maximum field of view: 0.9°
Magnitude limit: 13.3
Resolution limit: 0.7 arc second
Finder: 1× power red-dot
Objects in HC database: 18 000
Weight (no tripod): 22 lb (10 kg)

quality, but you will definitely be purchasing additional ones to exploit the potential of these telescopes.

Planetary views are outstanding; steady seeing will show wonderful detail on Jupiter, Saturn, and Mars. Even with the NexStar 5, you will see the ever-changing festoons, ovals, and other details in Jupiter's bands. With the NexStar 8, the amount of detail will be limited only by seeing conditions. The Great Red Spot and moon/shadow transits will also be a treat. Saturn will show off the Cassini Division and on exceptional nights the Crepe ring will be visible. In the NexStar 8 you will be able to sight bands on Saturn's disk. Favorable passes of Mars will show surface markings and the polar caps. Naturally, the Moon will not disappoint either.

The NexStar 5 will show you hundreds of Deep Sky Objects and detail in many. Spiral arms in galaxies, resolved stars in globular clusters, contours in gaseous nebula – all are visible on clear, dark nights. But many say that the deep sky really starts getting interesting in an 8-inch scope. If DSOs are your favorite target, seriously consider the NexStar 8, which has enough aperture to show much fainter objects, GoTo technology to put them in the field of view, and Celestron's outstanding optics – all at a price within reach of the serious amateur astronomer.

NexStar 5i and 8i

The NexStar 5i and 8i are Celestron's updated versions of the original NexStar 5 and 8. Optically and mechanically they are the same; the improvements are mostly in the electronics.

The differences are significant. First, the 5i (Figure 3.7) and 8i (Figure 3.8) are modular in design. Unlike all other current NexStars, the basic model does not include a computerized hand control. Instead they are supplied with a basic, electronic hand control that simply allows directional control of the motorized mount. Even with the basic hand control, the mount is capable of tracking the sky at sidereal rate when the telescope is mounted on an equatorial wedge. The optional $150 computerized hand control is Celestron part number 93962 and allows GoTo and automatic tracking of the moving sky.

The new mount now has two ports on the base. The first is the Auto Guide port for connecting a CCD autoguider. With the scope tracking in sidereal rate on a wedge (with the standard electronic hand control or the optional computerized hand control), an autoguider allows the scope to correct automatically for tracking imperfections during long-

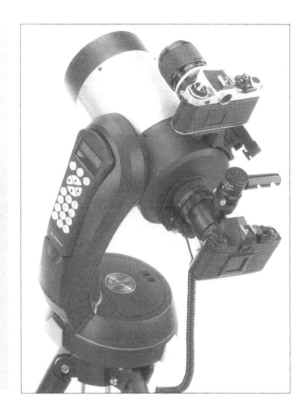

Figure 3.7. NexStar 5i. Photo courtesy Celestron.

exposure astrophotography. While the NexStar 5i and 8i are not the most precise mounts, this does allow for relatively inexpensive entry into astrophotography.

The second port is labeled AUX (Auxiliary).This is a more general-purpose connection, but the only device currently on the market to take advantage of it is Celestron's CN-16 GPS module (Figure 3.9). The $200 CN-16 module, Celestron part number 93963, includes a GPS receiver, electronic leveling sensor, and electronic compass. These three tools greatly simplify the initial alignment by providing all the necessary information for the telescope to locate the two required alignment stars.

However, you must of course have the computerized hand control to make use of alignment and GoTo technology. The optional $150 computerized hand control for the 5i/8i is the result of the evolution of the NexStar system from the initial NexStar 5 on up to the NexStar 8/11 GPS models. It includes all the features Celestron

NexStar 5i at a Glance

Approximate street price: $900 (no tripod)
Objects in optional computerized HC database: 40 000
Other specifications same as NexStar 5 above

NexStar 8i at a Glance

Approximate street price:$1,200 (no tripod)
Objects in optional computerized HC database: 40 000
Other specifications same as NexStar 8 above

Figure 3.8. NexStar 8i. Photo courtesy Celestron.

has developed to date. We will cover those features in later chapters, but here are some of the more exciting:

- Polar alignment (insuring the wedge is properly aligned with the North or South celestial pole) is extremely simplified. No more need for a "sight tube" or polar alignment finder on the wedge.
- Auto Alignment is now available when mounted on a wedge. This allows the scope to locate the two alignment stars automatically, while previous NexStar models required you to manually locate alignment stars when using a wedge.
- The Hibernate feature allows you to power off the scope and yet the star alignment is preserved. When you power up again, you can immediately resume observing. This is useful for telescopes permanently mounted in an observatory; you may not need to realign for weeks! Another use is observing planets in daylight. Align at night, hibernate the scope, then go out the next day for accurate GoTo performance in the daylight.
- Slew and Filter limits, as described previously for the new GT hand control for the little NexStars, are now available.

These new features are a significant improvement in the usability and "friendliness" and the NexStar 5i and 8i are still the most portable GoTo scopes in their class. Either model is highly recommended.

Figure 3.9. CN-16 GPS Module. Photo courtesy Celestron.

NexStar 8 GPS and 11 GPS

And finally, we arrive at the flagships of Celestron's NexStar line: the 8/11 GPS (Figures 3.10 and 3.11). Optically, the NexStar 8 GPS is identical to the NexStar 8, providing the same visual performance as described for the NexStar 8 earlier. The optics of the NexStar 11 are exquisite, consistently being described by owners and reviewers as the best optics they have seen in mass-produced Schmidt–Cassegrain telescopes.

If, as mentioned earlier, the deep sky starts to get interesting in an 8-inch scope, it is downright exciting in an 11-inch. The extra light grasp of the NexStar 11 brings thousands of additional deep sky objects into view. From dark, clear skies you will see incredible detail on galaxies, nebulae and star clusters. Even in light-polluted urban skies, globular clusters bring a "wow" to my lips time after time. Texture and dark lanes are readily apparent on many galaxy and nebula. If the deep sky thrills you, this is the NexStar you want.

And planetary detail will impress as well. A well-collimated NexStar 8 or 11 will show you as much detail and contrast as the seeing conditions will allow. Jupiter will show great detail and color in its many bands. The Great Red Spot shows a clear border from the South Equatorial Belt. Moon and shadow tran-

NexStar 8 GPS at a Glance

Optical system: Schmidt–Cassegrain catadioptric
Approximate street price: $2500 (includes tripod)
Aperture: 8 inches (203 mm)
Focal length: 2032 mm
Focal ratio: f/10
Supplied eyepiece: 40 mm Plössl (51×)
Maximum magnification: 400×
Maximum field of view: 0.9°
Magnitude limit: 13.3
Resolution limit: 0.7 arc second
Finder: 9×50 straight through
Objects in HC database: 40 000
Weight (no tripod): 42 lb (19 kg)
Tripod weight: 26 lb (11.8 kg)

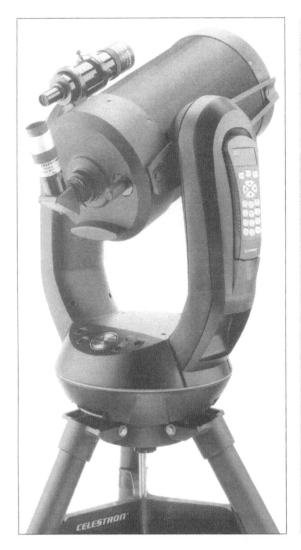

Figure 3.10. NexStar
8 GPS. Photo courtesy
Celestron.

sits across Jupiter are easily visible. Saturn shows clear banding as well as the Cassini Division and the Crepe ring. Saturn's shadow on its rings is easily visible; even the faint shadow of the rings on the planet disk is visible on exceptional nights. The five brighter moons of Saturn are readily visible in the NexStar 11. Mars will delight with easily visible detail and color as well the polar icecaps.

Unlike all other NexStars, the NexStar 8/11 GPS models sport more traditional, two-armed fork mounts. These scopes were built for extreme stability and astrophotography. Both axes are worm-driven, 5.6-inch bronze gears. The critical right ascension axis rides on a large, stable 9.5-inch roller bearing track. Power is supplied via an external 12-volt power supply – there is no internal battery option. The power supply plugs into the

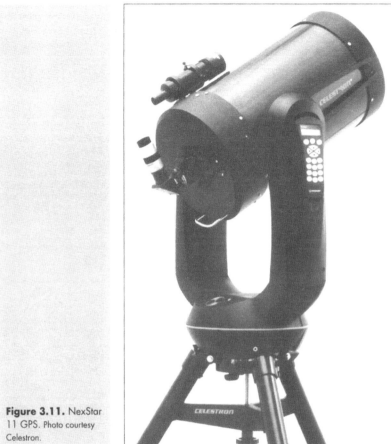

Figure 3.11. NexStar 11 GPS. Photo courtesy Celestron.

stationary section of the base and an internal slip-ring provides power to the hand control, internal electronics, and motors. This results in no power cable to tangle as the scope is moving from object to object. In fact, all models of NexStar allow for unlimited revolution in the azimuth/RA axis, as there is no internal hard stop. Both scopes come with a heavy-duty adjustable tripod and also include vibration suppression pads used under the tripod to dampen vibrations.

Another innovation for astrophotography is the carbon fiber tube material. Schmidt–Cassegrain scopes are very sensitive to small changes in focus. The folded light path makes every movement of the primary mirror – used to focus SCTs – the equivalent of five times as much movement in a refractor. This doesn't cause a problem when we are using the scope visually, but during a long photographic exposure, the ambient temperature can cause the optical tube to contract and throw off the focus. Carbon fiber has an expansion/contraction factor of one-fourth that of traditional aluminum tubes, largely eliminating this problem.

Both models have a dedicated autoguider port, allowing a CCD autoguider to keep the scope precisely centered on an object during a long-exposure photograph. Both also provide

Permanent Periodic Error Correction (PPEC) to smooth out any small errors in the right ascension worm gear drive.

The NexStar 8 GPS and 11 GPS both also feature Celestron's Fastar optical system. With the Fastar system you are able to remove the secondary mirror assembly and replace it with a lens and CCD camera. The focal ratio in Fastar configura-

NexStar 11 GPS at a Glance

Optical system: Schmidt-Cassegrain catadioptric
Approximate street price: $3000 (includes tripod)
Aperture: 11 inches (279 mm)
Focal length: 2800 mm
Focal ratio: f/10
Supplied eyepiece: 40 mm Plössl (70×)
Maximum magnification: 550×
Maximum field of view: 0.6°
Magnitude limit: 14
Resolution limit: 0.5 arc second
Finder: 9×50 straight through
Objects in HC database: 40 000
Weight (no tripod): 65 lb (29.5 kg)
Tripod weight: 26 lb (11.8 kg)

GPS Firmware Versions

The NexStar GPS scopes all have two sets of processors. There are processors in the hand control (HC) and there are processors in the base of the scope in a section known as the motor control (MC). The programs that these processors use to do their jobs are called firmware. In the GPS scopes there have been several versions of both.

MC Versions

1.0 – Shipped only in the original NexStar 11 GPS. This version had some problems with smooth tracking and should definitely be upgraded. The most detrimental problem is a vibration that affects high magnification views and astrophotography.

2.0 – Shipped for a short while in both the NexStar 8 GPS and 11 GPS. This version overcame problems with tracking, but there were still some problems with vibration in some telescopes.

3.0 – Shipped with many NexStar 8 GPS and 11 GPS scopes. This version is characterized by the "bee in a can" tracking sound that solved the vibration issue. It does not support some of the newest features such as cordwrap prevention and advanced control via a Personal Computer.

4.1 – This version features silent, accurate tracking and support for all the latest NexStar features. It also improves the encoder resolution to 0.112 arc second.

4.3 – The current version at the time of this writing. Incorporates a slightly improved tracking algorithm that eliminates a slight hop seen in some scopes.

HC Versions

1.2 – Shipped with original NexStar 8 GPS and 11 GPS scopes. Several features have since been added, and if you have this version you should definitely upgrade.

tion is f/2, allowing reduced exposure times – just $\frac{1}{25}$ the time required to image at the eyepiece port. This also allows you to use a smaller CCD chip to provide the same area of sky coverage. Cameras with smaller CCD chips are less expensive than their larger counterparts.

And then there are the features that give these scopes their name. The NexStar 8/11 GPS telescopes have a built-in GPS module, downstop switch (for detecting level), and electronic compass. The GPS module links with the Global Positioning System satellites to determine the precise time and location (longitude and latitude). Originally the NexStar 8/11 GPS shipped with a 12-channel GPS module, but it now comes with a 16-channel module. The new CN-16 GPS module for the NexStar 5i and 8i is a 16-channel model as

GPS Firmware Versions *(continued)*

1.6 – The updated hand control with all features described in this book. Shipped with NexStar 8 GPS and 11 GPS scopes.
2.2 – A minor upgrade that includes support for the NexStar 5i/8i and a slightly more accurate Wedge Align routine.

Determining Versions

With hand control version 1.6 or higher, you will find a Version option on the Utility menu. See Chapter 5 for more details. With hand control version 1.2, things are a little more complicated as there is no Version command. Basically, if there is no Version command on the Utility menu, you need an upgraded hand control. Customers in the United States should contact Celestron via Technical Support at their web site – www.celestron.com. Customers in other countries should first contact their Celestron dealer.

well. In practice, this provides no additional advantage, as each channel is capable of linking with just one satellite at a time. To determine precise time only requires a link to one satellite, to determine precise longitude and latitude only requires linking to three satellites. Additional channels and satellites allow for determining altitude above sea level, which is of no practical use in alignment of the telescope.

The downstop switch is used to position the telescope at a position perpendicular to the azimuth axis in order to provide a starting point for the encoder system that provides GoTo positioning. The downstop switch is also used to position the telescope parallel to the fork arms when performing a wedge alignment.

The electronic compass points the telescope at magnetic north. This varies by differing amounts at various locations on Earth, so the latest version of the NexStar 8/11 GPS hand control includes a routine to calculate and store the offset for future use. Pointing north is required for the telescope to point automatically at the two stars during the initial alignment procedure. Combined, the GPS, downstop switch, and electronic compass allow you to turn on the scope and, faster than you can set out your eyepieces, it will be waiting for you to center the first alignment star.

If it seems that the NexStar 8 GPS and 11 GPS are made for you, do take into account their substantial weight. After several trips carrying my NexStar 11 up and down a flight of stairs in my home I sometimes consider it to be exercise equipment rather than astronomical equipment!

CGE Series

Although Celestron does not bill the CGE series as NexStar telescopes, they do use the NexStar computerized hand control to provide GoTo performance, so I will discuss them here. The CGE series consists of four different models, all featuring Schmidt–Cassegrain optical tubes. Celestron offers the CGE in 8-, 9.25-, 11-, and 14-inch apertures, with model numbers of CGE 800, 925, 1100, and 1400.

Unlike the NexStars, CGE scopes are not provided on a fork mount, but are based rather on a German equatorial mount (GEM). The CGE GoTo mount is also sold separately, for about $3000, allowing freedom to choose the optical tube most suited to your needs. GEM mounts are more portable than twin fork scopes and provide better balance and more accurate tracking for extremely high-performance astrophotography.

The optical tube can be easily removed from the mount for transport, a feature not found in fork-mounted scopes. This allows the CGE 1400 (Figure 3.12) to be the most

Figure 3.12. CGE 1400. Photo courtesy Celestron.

portable 14-inch GoTo scope ever produced. And the German EQ mount is equally at home permanently housed in an observatory.

The 14-inch optical tube makes many faint deep sky objects positively bright. Globular clusters, nebulae, and galaxies will show incredible detail, as will the Moon and planets. Be careful about taking a look through the CGE 1400 – you might not be able to go back to your old telescope! Perhaps the only things that will temper your enthusiasm are the price and weight. While the CGE 1400 is portable, you won't likely fit it into a midsize car and setup is best executed by two persons. And the price is definitely in the range of serious commitment.

The 8- and 11-inch models share the same optical tube as the NexStar 8 GPS and 11 GPS and perform as previously described. Celestron's 9.25-inch is a unique design in SCTs with a slightly different geometry. Many who have looked through it consider it the finest SCT on the market, particularly for viewing planets. With light-gathering ability increased over

CGE Series at a Glance

	CGE 800	CGE 925	CGE 1100	CGE 1400
Cost	$3500	$3800	$4150	$5800
Aperture	8" (203 mm)	9.25" (235 mm)	11" (279 mm)	14" (355.6 mm)
Focal length	2032 mm	2350 mm	2800 mm	3910 mm
Focal ratio	f/10	f/10	f/10	f/11
Eyepiece	25 mm Plössl	25 mm Plössl	40 mm Plössl	40 mm, 2" barrel
Max. magnification	400×	470×	550×	710×
Max. FOV	0.9°	0.74°	0.6°	0.56°
Magnitude limit	13.3	13.5	14	14.6
Resolution limit	0.7 arcsec	0.6 arcsec	0.5 arcsec	0.4 arcsec
Finder	6×30	6×30	9×50	9×50
Optical tube	Carbon fiber	Aluminum	Carbon fiber	Aluminum
Fastar	Yes	No	Yes	Yes
Optical tube weight	12.5 lb(5.7 kg)	20 lb (9.1 kg)	27.5 lb (12.5 kg)	45 lb (20.5 kg)
Equatorial mount weight		42 lb (19.1 kg)		
Tripod weight		41.5 lb (18.9 kg)		

the 8-inch models, slightly fainter objects are within reach. Also, a slightly improved resolving capability is realized.

Celestron has included a host of features designed with serious astrophotography in mind. The mount includes autoguide capability and Permanent Periodic Error Correction, as described above for the NexStar 8 GPS and 11 GPS. The drive gears are large-diameter worm-driven gears for low periodic error and smooth, responsive tracking and guiding. The optical tube of the 8-, 11-, and 14-inch models features Celestron's Fastar option to allow faster CCD imaging at f/2 focal ratio. The optical tube on the 8- and 11-inch models is constructed of carbon fiber for improved thermal expansion characteristics. We should expect the 9.25- and 14-inch models to be upgraded to carbon fiber in the future.

The CN-16 GPS module also works with the CGE mount, providing added convenience during alignment and extremely accurate timing of celestial events. Even if you don't opt for the GPS module, the mount includes a real-time clock/calendar. The tripod is an improved design to provide added stability for heavier loads.

At the time of this writing the CGE series is too new to include further information and is not discussed in later chapters. The hand control includes the standard NexStar features, so many of the topics discussed later in the book apply.

Which Scope for You?

No single telescope is perfect in every situation. In deciding what to buy, there are many factors you must take into account. In deciding on just one scope, there will always be tradeoffs. One scope has the aperture you need for the faint objects you are after, but its narrow field of view doesn't do justice to some of the sky's wide showpieces. Another scope is highly portable, but doesn't give much detail on the planets. Yet another scope has exquisite optics, but the price tag is out of your reach. So we must decide which scope best meets our needs.

The aperture of a scope determines, more than any other factor, what objects you will be able to see. If faint fuzzies (deep sky objects) are your favorite targets, then 8 inches is probably your minimum aperture. The NexStar 8, 8i, 8 GPS, 11 GPS, and CGE series will all give superb views of the deep sky. The smaller NexStars will show you many deep sky wonders as well, but with limited details. Aperture will also determine how much detail you will see on smaller objects such as planets and lunar craters, so ultimately you want to buy the largest aperture you deem practical and affordable.

Portability might be an issue. If you must move the scope frequently or far, a 65-pound scope will likely see little use. If travel is required to reach your favorite observing site, you must naturally consider the size of a telescope and if it will fit into your vehicle. Any of the little NexStars – the 60, 80, 114 and 4 – are very good travelers. The NexStar 5, 5i, 8, and 8i are the most portable scopes on the market in their aperture ranges. And even the behemoth NexStar 11 GPS is quite portable when we consider telescopes capable of precision astrophotography in the 11-inch range.

You should also consider the angular size of your targets. Some objects, such as large nebulae and open clusters, require a wide field of view. The longer the focal length of a telescope, the narrower the maximum field of view. The Andromeda Galaxy (M31) is 3° across; the only NexStar capable of taking in that sight is the NexStar 80. The Orion Nebula (M42) is about 85 arc minutes across – or $1\frac{1}{3}$°. The 8-inch NexStars cannot encompass this in their field of view with 1.25-inch eyepieces and the NexStar 11 cannot reach this field of view even with 2-inch eyepieces.

If astrophotography seriously interests you, be certain to buy a telescope well suited to the task. The little NexStars and the NexStar 5 and 8 do not track well enough for long-exposure astrophotography. And none of them allow for the use of an autoguider. The NexStar 5i and 8i do provide for the use of an autoguider, but they do not have extremely accurate tracking. Results might be good, but not perfect. The 8 GPS, 11 GPS, and CGE series are the astrophotography workhorses of the NexStar line and are fine visual-use scopes as well. But their added precision comes at a cost, both monetarily and in size and weight.

Of course, we should not ignore our budget. Paying the rent and eating should generally take precedence over new astronomy equipment; after all, your scope needs a dry roof over its head! Set a limit on what you will spend before you start. You may decide to exceed this limit as you learn more, but at least you will have an expectation of what models of telescopes are within your reach and you will realize when you need to be patient and save for a while longer.

Whichever telescope you choose, NexStar or otherwise, there is a whole new experience awaiting you.

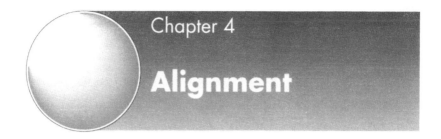

Chapter 4

Alignment

Much of the material in this chapter is based upon the "NexStar Alt–az Alignment Guide" written by John Carlyle, Alain Fraysse, Carroll Morgan, and myself in February 2001. The original document is available in the files section of the Yahoo NexStar Group, as well as at the NexStar Resource Site. Additional insight on the NexStar 8i was provided in tests performed by Jean Piquette and presented in his "Alignment Tips for i Series Telescopes", also available in the files section of the Yahoo NexStar Group.

Aligning and Using the Star Pointer or Finder Scope

Before you start your first alignment, you must align the Star Pointer or finder scope with the main telescope. This is best done during the day.

Looking through the main scope with a moderately powered eyepiece, aim your telescope at a distant object such as a ceramic insulator on an electric pole. To align the Star Pointer provided with most NexStar models, turn the unit on and then use the two knobs (one at the front, the other at the back) to center the red dot over distant object. For the finder scope provided with the NexStar 8/11 GPS, use the three thumbscrews to center the crosshairs on the object. Check to be certain that the finder is well secured and you are finished. At night, you might want to fine-tune on a bright star, after your scope is tracking the sky.

When using the Star Pointer you might find that even at the lowest setting it is too bright. Try switching the power on and off while centering the star. You can also substitute a 1.5-volt battery centered in a rubber washer in place of the standard 3-volt battery.

In order for a NexStar telescope to GoTo objects and track them, it is necessary for the computerized hand control to build a model of the sky. An accurate model can be built if we point the scope at two known stars, a process we refer to as initial alignment, or, simply, alignment. While even a haphazard approach to alignment will provide fairly accurate GoTo and tracking, there are several key elements to producing very accurate performance, night after night.

As I mentioned in Chapter 3, this book generally describes the behavior of the latest versions of the NexStar firmware (internal program). If your telescope does not have a particular feature or does not behave as described, first check to be sure I indicated that this feature is available for your model of telescope. If so, refer to Chapter 3 to determine what version firmware you have and whether you should contact Celestron about an upgrade.

Must Do's of Alignment

The most critical requirements for an accurate alignment are as follows:

- For the older NexStars, start the alignment procedure with the optical and azimuth axes perpendicular.
- Change the altitude GoTo Approach to Positive for NexStar 8/11 GPS and NexStar 8i telescopes.
- When in the Southern Hemisphere, change the azimuth GoTo Approach to Negative for NexStar 8/11 GPS and NexStar 5i/8i telescopes.
- Approach the perpendicular or index position from the correct direction.
- Approach north or the starting direction from the correct direction.
- Accurately center the alignment stars.
- Center the alignment stars with the correct final movement.
- Choose stars with good separation and best placement in the sky.

These are so significant we will discuss each in turn.

For the NexStar 5/8 and NexStar 60/80/114/4 Start the Alignment Procedure with the Optical and Azimuth Axes Perpendicular

The older NexStar models are sensitive to very small differences in the position of the optical tube when the alignment procedure starts. For these models, it is critical to start the alignment with the optical axis and the azimuth axis perpendicular. The optical axis is basically the centerline of the optical tube. Since all models of NexStar scopes have straight tube walls, the outside of the optical tube is an acceptable substitute for the optical axis. The azimuth axis comes straight up from the center of the mount. Auto-Alignment, Two Star Alignment, and Quick Alignment all require that you point the telescope north and "level", but actually what is needed is to point north and set the optical and azimuth axes perpendicular.

There are two ways to insure you start perpendicular. The first is to set up the tripod, without the mount and scope, and accurately level the top of the tripod. Then mount the scope and accurately level the scope. The more accurate you are at this two-step leveling, the better the GoTo performance you will receive. This will require several minutes each time you set up the scope.

A simpler and faster method also proves to be more accurate. In the comfort of your home, set up the tripod and level it precisely, rotating the level around the head of the tripod to insure accuracy. Then mount the scope on the tripod and carefully level it, using the up and down direction buttons on the hand control to make adjustments. Use a rate of 4 or lower to make small movements. As shown in Figure 4.1, place a piece of tape (I like to use masking tape) on both sides of the joint where the fork arm meets the optical tube holder. Then use a short ruler and mark a straight line across both pieces of tape.

From that point on, when you set up the scope, you will not even need to remove the scope from the tripod, just roughly level the tripod. Note that the more accurately you level the tripod, the closer the scope will be when initially pointing at the two alignment stars. When the hand control prompts for "north and level", aim the scope towards the north and line up the index marks on the fork arm.

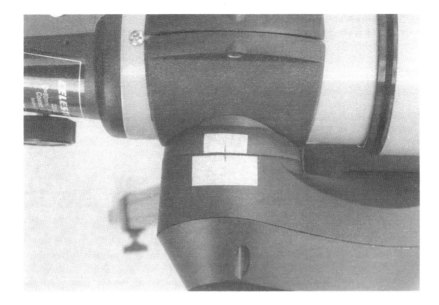

Figure 4.1. Tape to mark the location where the optical tube is perpendicular to the azimuth axis.

Change the Altitude GoTo Approach to Positive for Nexstar 8/11 GPS and Nexstar 8i Telescopes

Change the altitude GoTo Approach to **Positive** on the NexStar 8/11 GPS and the NexStar 8i to take advantage of the default final movements of the GPS Alignment routine and the natural back-heavy balance of these scopes. GoTo Approach is found on the Scope Setup menu. Both axes – altitude and azimuth – can be set to either Positive or Negative. The default is altitude Negative. In altitude, Positive means the final approach will be with the front of the scope moving up as viewed from the back of the scope. This setting is permanent and as such is only required once.

If you do not change the altitude approach, you must make all final movements during alignment to move the front of the scope down, rather than up as I will recommend later.

When in the Southern Hemisphere, Change the Azimuth GoTo Approach to Negative for NexStar 8/11 GPS and NexStar 5i/8i Telescopes

The majority of celestial objects observed by those in the Southern Hemisphere are located to the observer's north. Such objects are moving to the scope's left, as are all objects when mounted on a wedge. An azimuth GoTo Approach of Negative causes the final GoTo motion to be towards the left, starting tracking immediately without delay caused by gear backlash. GoTo Approach is found on the Scope Setup menu and the default setting is Positive. This setting is permanent and as such is only required once.

If you do not change the azimuth approach, you must make all final movements during alignment to move the front of the scope right, rather than left as I will recommend later.

Approach the Perpendicular or Index Position from the Correct Direction

In order to minimize the effects of gear backlash on the altitude index position, it is best to approach your index or perpendicular mark while traveling the correct direction. This varies from model to model:

- **NexStar 60/80/114/4** Move the front of the scope UP to your index marks. Be precise.
- **NexStar 5/5i/8** Move the front of the scope DOWN to your index marks. With the NexStar 5/8, be precise. With the NexStar 5i there is no need to be too concerned with accuracy when aligning the index marks; just be sure your final approach is in the downward direction.
- **NexStar 8/11GPS and the NexStar 8i** Move the front of the scope UP to your index marks. There is no need to be overly concerned with accuracy when aligning the index marks, just be sure your final approach is in the upward direction.

With the older NexStars, it is important to accurately align the index marks. If you over-shoot the index mark, move back, well past the mark, and approach again more carefully from the correct direction. You may find it easier to align the marks at rate 6 or 7.

Approach the North or the Starting Position from the Correct Direction

In order to minimize the effects of gear backlash on the azimuth index position, it is best to approach North – or simply the starting position for a Two Star Alignment – with a final movement in the correct direction. Approach north by pressing the RIGHT arrow button. In the Southern Hemisphere, if you are using a NexStar 8/11 GPS or NexStar 5i/8i, approach north by pressing the LEFT arrow button.

Accurately Center Alignment stars

Eyepiece selection when centering alignment stars will be the most critical factor in this case. If you align the stars using a wide-field, low-magnification eyepiece then you can expect that GoTo will place objects in the field of view of that eyepiece, but perhaps outside of a higher magnification eyepiece with its much smaller field of view. Thus, it is best to use your highest magnification when centering the two alignment stars. Use a Barlow lens for even greater accuracy. You will likely find it necessary to switch to a lower power first to insure you have the right star centered and then switch to the higher power for final centering.

Placing the alignment stars in the center of the field of view can be challenging. The telescope is not tracking the motion of the sky at this point and so the star drifts quickly in a highly magnified view. You may find it necessary to set the star off-center and let it drift, pressing the Align button at the correct moment. Additionally, centering a tiny point of light in a barely visible circle can be a challenge. Special eyepieces known as reticle eyepieces can be used to provide a crosshair in the field of view. Another method is defocusing the star, as this transforms the star into a large disk that is much easier to center.

Center Alignment Stars with the Correct Final Movement

To help minimize the effect of backlash or play in the gears, your final movements when centering the alignment stars are essential. The correct direction varies from one model of NexStar to another:

- **NexStar 60/80/4** Center the stars with the RIGHT and DOWN arrow buttons (this assumes the new GT hand control – for the original GT hand control, use RIGHT and UP).
- **NexStar 114** Center the stars with the RIGHT and UP arrow buttons.
- **NexStar 5/5i/8** Center the stars with the RIGHT and UP arrow buttons.
- **NexStar 8/11GPS, NexStar 8i** Center the stars with the RIGHT and DOWN arrow buttons.

One caveat: if you have a NexStar 8/11 GPS or NexStar 5i/8i, use the LEFT arrow button rather than RIGHT if you are in the Southern Hemisphere.

It should be noted that the NexStar 5/5i/8 must be front-heavy for the best GoTo accuracy. The NexStar 8 is normally back-heavy and the NexStar 5/5i becomes back-heavy if you use very heavy eyepieces or mount a camera off the back. To remedy this, add weight to the front or consider a set of brackets as discussed in Chapter 13.

If you overshoot your target, move back, well past the target, and approach more carefully from the correct direction.

Arrow Buttons for Final Approach to Starting Position and When Centering Stars

	Northern Hemisphere		
	Index mark or perpendicular position	North or two-star align starting position	Centering stars
N60/80/4 with New HC	UP	RIGHT	RIGHT and DOWN
N60/80/4 with Old HC	UP	RIGHT	RIGHT and UP
N114	UP	RIGHT	RIGHT and UP
N5/5i/8	DOWN	RIGHT	RIGHT and UP
N8/11GPS and N8i *	UP	RIGHT	RIGHT and DOWN

	Southern Hemisphere		
	Index mark or perpendicular position	North or two-star align starting position	Centering stars
N60/80/4 with New HC	Same as above for all movements.		
N60/80/4 with Old HC	Same as above for all movements.		
N114	Same as above for all movements.		
ˉN5/8	Same as above for all movements.		
N5i †	DOWN	LEFT	LEFT and UP
N8/11GPS and N8i * †	UP	LEFT	LEFT and DOWN

Notes:
*The UP/DOWN movements indicated for the N8/11GPS and N8i are dependent upon changing the altitude GoTo Approach to Positive as discussed earlier.
† The LEFT movements indicated for the N8/11GPS and N5i/8i are dependent upon changing the azimuth GoTo Approach to Negative as discussed earlier.

Choose Stars with Good Separation and Best Placement in the Sky

When using the Two Star Alignment method, you must select the two alignment stars. For Auto Alignment and GPS Alignment, the telescope chooses the two alignment stars for you. In any case, there are a couple of general rules to keep in mind in choosing the alignment stars. If the Auto Alignment or GPS Alignment chooses poor stars, pressing the Undo button will cause the scope to move on to the next potential star in the list. If the scope is moving towards an undesirable star, you do not need to wait until it arrives at its target; press any of the arrow buttons on the hand control to cancel the slew, then press Undo to move to the next star.

Selection of stars varies slightly depending upon whether you are using the scope in alt–azimuth mode (mounted directly on the tripod) or equatorial mode (mounted on a wedge). Here are the main selection criteria:

- In alt–az mode, if you are in the Northern Hemisphere, using Polaris as an alignment star generally improves GoTo performance. However, on a wedge, it is important to avoid using Polaris as an alignment star.

- Second, choose alignment stars that are well separated. I recommend a separation of between 90° and 135°. In general, consider your choices and choose the two that are farthest apart.

- Avoid alignment stars below 20° in altitude (angle above the horizon). Stars too close to the horizon are displaced optically by the effect of refraction through the Earth's atmosphere.

- In alt–az mode, avoid stars above 70° in altitude. Aligning on stars too high in altitude exaggerates inaccuracy in centering the star due to the high number of motor turns required to move small increments in azimuth. The equivalent rule when using a wedge would be to avoid stars above 70° or –70° declination, but other than Polaris (already ruled out above) none of the alignment stars falls in that range.

Alignment Methods

The various models of NexStar scopes provide several methods for alignment; we will discuss each separately. The manual lays out fairly easy-to-follow, step-by-step procedures; I will use a narrative format to emphasize important details. The designations in parentheses indicate which models provide each method. For information about various versions of hand controls, refer to Chapter 3. Button and menu names are shown in **bold** print.

GPS Alignment

(NexStar 8/11 GPS, NexStar 5i/8i with optional GPS module)

The simplest and most accurate alt–az alignment method is GPS Alignment. The Global Positioning System allows for very accurate determination of longitude, latitude, date, and time to provide much of the information required to automatically point towards the two alignment stars. In addition the telescope needs the direction of north and the location of the point where the optical tube is level. An electronic compass provides north and level is determined by an internal switch (known as the downstop switch on the NexStar 8/11 GPS) indicating the point where the optical tube is approximately level.

To perform the GPS Alignment, set up the tripod and roughly level it. The closer you are to leveling the tripod, the closer the scope will be to locating the alignment stars. Mount the scope on the tripod, point roughly north and level, and on the NexStar 8/11 GPS, engage the clutches. Don't overlook the azimuth clutch; its location under the base encourages forgetfulness. Insert a high-power eyepiece, turn on the scope, and select **GPS Align** to start the procedure.

The telescope will dip slightly then return to level, stopping when it reaches level. On the NexStar 8/11 GPS, the downstop switch will make an audible click. It will then swing left and right, zeroing in on magnetic north. After finding north, the hand control will show the local time

Initial Setup of the GPS Model Hand Control

Calibrating the Compass and the Level

The internal compass and downstop switch ("level" or "altitude" sensor) on the NexStar 8/11 GPS and the NexStar 5i/8i with optional GPS module are roughly calibrated at the factory. Both should be calibrated for improved accuracy of the GPS method in locating the two alignment stars. Calibrating the compass is necessary to compensate for the difference between magnetic and true north at your location, allowing the scope to come much closer when initially pointing at the two alignment stars. Calibrating level is required to determine the point where the optical tube is perpendicular to the fork arms or close to level.

- Start by very carefully leveling the tripod in all directions prior to attaching the scope. Set the level on the head of the tripod, testing alternately across the top of two legs at a time. Adjust the legs until level in all orientations.
- Next perform a GPS alignment. Take care to accurately center the alignment stars and, after completion, check to be sure that GoTo is accurately placing objects in the field of view.
- Press the Menu button, select Utilities, Compass, and then Calibrate.
- On the Menu button, select Utilities, Level (or Alt Sensor), and then Calibrate.

Note that even after calibrating the compass and level, during alignment, the scope will still point off from true north and a true perpendicular position. The calibration routines do not change the positioning of the internal compass or downstop switch, rather they store compensation values used in future alignments.

If you travel a significant distance with your scope, the new location may have a different magnetic north deviation and thus require the compass to be recalibrated. You should not need to recalibrate level unless you reset the hand control to factory defaults or return your hand control to Celestron for servicing.

Setting Local Time Zone

Although the GPS system provides the hand control with time accurate to the millisecond, it does not provide your local time zone. When shipped from the factory, each hand control is set to Pacific Standard Time. If you wish to use the View Time-Site feature to display current local time, you will need to insure the time zone and daylight savings time settings are correct.

Press the Menu button, and select Scope Setup, Setup Time-Site. Enter past the current time, and then choose either Standard Time or Daylight Saving as appropriate. Next select the correct time zone and enter through the date, longitude and latitude. If your location uses daylight savings time, you will need to repeat this procedure each spring and fall, otherwise you should not need to revisit this menu item unless you reset the hand control to factory defaults or return your hand control to Celestron for servicing.

for the last time you used the scope and the top line of the display will cycle through the messages "GPS Searching", "Enter if OK", and "UNDO to edit". The GPS receiver will shortly lock onto three GPS satellites and update the date, time, and location and then slew to the first alignment star. If large buildings, hills, or other obstructions block any of your horizons, the GPS link could take some time. Also, if you have transported the scope more than a few hundred miles, additional time will be necessary to download database information from the satellites. In any case, if you become impatient waiting for a GPS link, you can use the **Up** and **Down** buttons (**6** and **9** on the keypad) to scroll through the date, time, and location settings. If they are all correct (not very likely), press **Enter** to accept and proceed with the alignment. If they need adjusted, press **Undo** and make any necessary corrections.

At this point, the telescope will slew (point) towards the first alignment star. If you know that the first star is behind an obstruction, or if you do not feel it is a good choice, press any of the arrow buttons to cancel the slew, then press the **Undo** button to go to the next potential alignment star. After the scope has slewed to the general location of the first star, follow the directions on the hand control to center the star in the finder and press **Enter**. Your goal at this point is simply to get close enough to the star so that it is visible in the eyepiece. After pressing **Enter**, follow the directions on the hand control to center the star in the eyepiece and press the **Align** button. Repeat this process for the second alignment star. If you centered the correct stars, the hand control will report success; if not, you will be informed that the alignment failed and you will need to start over from the beginning.

If you follow the advice at the beginning of this chapter regarding alignment star selection, accurate centering in a high-power eyepiece and centering the alignment stars with the correct arrow buttons, your scope should place objects in the field of view all night.

Auto Alignment

(All models)

The next-easiest alt–az alignment method is Auto Alignment. You provide the longitude, latitude (or you can select a nearby city in some models of NexStar hand controls), date, and time information. You must also point the telescope north and set the optical axis perpendicular to the azimuth axis. The scope points near the two alignment stars, which you center, and you are finished.

To perform an Auto Alignment, set up the tripod and roughly level it. The closer you are to leveling the tripod, the closer the scope will be to locating the alignment stars. Insert a high-power eyepiece, turn on the scope, and select **Auto Align** to start the procedure.

When prompted to point north and level, if you can see Polaris (the North Star), use the arrow buttons on the hand control to center it in the finder. If you are relying on a compass, point as close to true north as you can. It should be noted that magnetic north and true north vary by as much as 20° in some locations, so compensate if you are aware of your local deviation. Regardless, pointing north need only be approximate; the telescope uses this information simply to locate the general location of the two alignment stars. Accurately centering the alignment stars is what is required to build the model of the sky. After pointing north, use the appropriate up or down arrow button to move the scope until your marks on the fork arm are aligned. For the older NexStar models, be as precise as possible and then press **Enter**.

At this point on the GPS models you will find that a GPS link may be established, providing location, date and time information. On all other models you must provide such information. Be aware that when you enter the date, you must use mm/dd/yy (US format); the most common error in Auto Alignment is using an incorrect date format.

Determining and Entering Your Longitude and Latitude

The GT hand control on the little NexStars allows you to select your location as one of hundreds of cities in its internal database or by entering your longitude and latitude. All other models require longitude and latitude. With the GPS models, the GPS receiver will provide this information. For other models, or on the rare occasion when you cannot obtain a GPS link, you must provide your location.

While you do not need extreme precision, as accuracy within a few degrees is more than adequate, a handheld GPS receive is probably the best method for determining longitude and latitude. In lieu of GPS, refer to a map or turn to the Internet. Most any map site will provide longitude and latitude, but I recommend the Heavens Above web site - http://www.heavens-above.com/countries.asp.

Keep in mind that North America, South America, most of England and Spain, the western parts of France and Africa are west longitude, while most of the rest of the world, Asia, Australia, India, the Middle East and most of Europe and Africa, are east longitude.

If your version of NexStar asks for + or - for latitude, + is north latitude. It is not generally worth the trouble to enter accuracy beyond minutes of longitude and latitude, so if your scope prompts for seconds (a third pair of numbers for longitude/latitude), round to the nearest minute and enter 00 for seconds. For example, enter 127° 44′ 00″ for a longitude of 127° 44′ 20″ or, as reported by some map sites, the decimal equivalent of 127° 44.33′.

Note that on the original NexStar 5 and 8, storing your longitude and latitude in adjacent "stored locations" will corrupt the entries. This results in the Auto Alignment procedure missing the two alignment stars. To prevent this problem, only store and use longitude and latitude in locations 0, 4, 8 or 1, 5, 9.

Correct entries can be confirmed, rather than retyped, by simply pressing the Enter button. Use the **Undo** button like backspace on the computer to correct mistaken key presses.

It is not necessary to be extremely accurate in providing location and time information. Selecting a city or using longitude and latitude that are even a few hundred miles away will not affect GoTo performance at all. Again, accurately centering the alignment stars is what is required to build the model of the sky. However, if you are unfamiliar with the location of the alignment stars, it may make it more difficult to determine which is the correct star to center. Inaccurate date and time information will have a similar effect, but will also affect GoTo accuracy on the planets and, more so, the Moon. Accurate date and time are necessary for the hand control to calculate the current location of these wandering objects. Planets move only slightly from day to day, but the Moon moves dramatically in just one hour.

After providing the initial information, the telescope will slew (point) towards the first alignment star. If you know that the first star is behind an obstruction, or if you do not feel it is a good choice, press any of the arrow buttons to cancel the slew, then press the **Undo** button to go to the next potential alignment star. After the scope has slewed to the general location of the first star, follow the directions on the hand control to center the star in the finder and press **Enter**. Your goal at this point is simply to get close enough to the star so that it is visible in the eyepiece. After pressing **Enter**, follow the directions on the hand control to center the star in the eyepiece and press the **Align** button. Repeat this process for the second alignment star. If you centered the correct stars, the hand control will report success; if not, you will be informed that the alignment failed and you will need to start over from the beginning.

If you follow the advice at the beginning of this chapter regarding starting perpendicular, correct approach to the index mark, alignment star selection, accurate centering in a high-power eyepiece, and approaching the alignment stars from the appropriate direction, your scope should place objects in the field of view all night.

Two Star Alignment

(All models except Original GT)

Perhaps the quickest alignment method, and certainly the one which gives you the most control over alignment star selection, is Two Star Alignment. In this method, you do not need to provide your location, date, or time information; you simply point at two known stars yourself and center them. Naturally the disadvantage for beginners is that you must know the location of the two stars in the sky.

To perform a Two Star Alignment, set up the tripod and roughly level it. Insert a high-power eyepiece, turn on the scope, and select **Two Star Align** to start the procedure. When prompted to level the tube, use the appropriate up or down arrow button to move the scope until the marks on the fork arm are aligned. For the older NexStar models, be as precise as possible and then press **Enter**.

Next you will be prompted to select the first alignment star from the list of 40 NexStar alignment stars. Use the **Up** and **Down** buttons (**6** and **9** on the keypad) to scroll through the list and press **Enter** to select your first star. Two Star Alignment defaults to a relatively slow speed for initial movement, so you might want to press the **Rate** button and then press **9** to step up to maximum speed. Follow the directions on the hand control to center the star in the finder and press **Enter**. Your goal at this point is simply to get close enough to the star for it to be visible in the eyepiece. After pressing

Common Problems with Alignment on the NexStar 60/80/114/4

This list summarizes several potential problems that can prevent you from aligning your NexStar 60/80/114/4, particularly if you have the original GT hand control or a Tasco StarGuide.

- Your NexStar 60, 80, or 114 might be mounted backwards in the tube ring. The tube ring should be oriented such that the bolt that holds it closed is on the top. Then, when facing the side where the hand control attaches to the fork arm, the front of the optical tube should be pointed to the right. Note that on the N114, the front of the optical tube is the end with the opening. Refer to your manual; the pictures are correct, even if it didn't come that way from the factory!

- If you own the original GT hand control, do not select the model of scope from the menu in the GT control; if you do then your scope cannot be aligned properly. Incorrectly setting the model will result in the scope missing the alignment stars by a significant, but consistent amount. Then after you manually center the stars, all GoTo attempts will miss by a significant but consistent amount. Even if you go back to one of the alignment stars the scope will miss. Refer to Select Model in Chapter 5 for the correct procedure to set the model to Custom.

- If you have the original GT hand control, do not store and use locations (longitude/latitude) if you are in the Eastern Hemisphere – Europe, Asia, Australia, etc. There is a problem with the GT control that prevents using stored locations with east latitude.

- Be sure you are entering the date in mm/dd/yy format.

- If you have the original GT hand control, don't use military (24-hour) time, the alignment will fail. You must use the 12-hour clock and set p.m./a.m. On all models, if you are aligning after midnight, entering 01:30 will still require you to change the p.m. to a.m.

- Watch for double key entries – they are common with the original GT control and can cause problems, particularly when entering longitude and latitude.

- If you are not in North America, you will need to select a time zone by number. In the original GT control, these are not the usual offset from UTC (GMT), but instead you should refer to the time zone charts in the manual. These equate to 24 minus the offset from UTC.

Common Problems with Alignment on the NexStar 60/80/114/4 *(continued)*

- If you are new to astronomy, it will be important that your scope points accurately to the two alignments stars, otherwise you might center on an incorrect star. To assure accurate pointing during Auto Align, level the tripod, center Polaris (the North Star) in the finder, and then lower the optical tube to the level position (it is best to use a set of marks as described in this chapter). Refer to the charts in the back of the NexStar instruction manual so that you won't be fooled into selecting an incorrect star.

- Occasionally the alignment procedure will be completely off-track. If your scope doesn't point anywhere near what you think is the correct star, you are best off pulling the plug and starting over.

Enter, follow the directions on the hand control to center the star in the eyepiece and press the **Align** button. If you find the scope is moving too fast to allow accurate centering in the eyepiece, press the **Rate** button and try 5 or 4 as a more reasonable rate. Repeat this process for the second alignment star. If you centered the correct stars, the hand control will report success; if not, you will be informed that the alignment failed and you will need to start over from the beginning.

If you follow the advice at the beginning of this chapter regarding starting perpendicular, correct approach to the index mark, alignment star selection, accurate centering in a high-power eyepiece, and approaching the alignment stars from the appropriate direction, your scope should place objects in the field of view all night.

As noted in the Auto Alignment section above, the location of planets and the Moon is dependent upon the current date and time. If you want to GoTo a planet, the date and time must be provided to the scope. On the GT models, go the **Utilities** menu and select **Setup Time-Site** to provide correct date and time information. This is simply known as **Date and Time** on the original NexStar 5 and 8, but these models will automatically prompt you for date and time prior to slewing to a solar system object. On the latest GPS hand control models, use **View Time-Site** on the **Menu** button as this will attempt to link to the GPS system. Allow the GPS to update the information, or access the **Scope Setup** menu, where you will find **Setup Time-Site** for manual entry.

Quick Alignment

(NexStar 8/11 GPS, NexStar 5i/8i, New GT)

Quick Alignment is provided as a way to get an approximate alignment during the day or when simply testing your scope indoors. This would allow testing of a PC control connection or reasonably accurate tracking of the Sun for solar observations. (**Note:** Read the information in Chapter 8 regarding filters for safe viewing of the Sun!) You provide the longitude, latitude, date, and time information. You must also point the telescope north and set the optical axis perpendicular to the azimuth axis. Centering stars is not required; thus the alignment is just approximate.

Note that on the GPS models, a GPS Alignment will be more accurate than a Quick Alignment due to the compass calibration that provides a fairly accurate fix on true north. Simply accept the positions the scope points out for the two alignment stars and tracking will begin.

To perform a Quick Alignment, set up the tripod and level it. The more accurately you level the tripod, the more accurate tracking will be. Turn on the scope, and select **Quick Align** to start the procedure. When prompted to point north and level, use a compass, or if you are setting up in your usual nighttime location perhaps there is a distant tree that you know to be due north. After pointing north, use the UP/DOWN arrow buttons to move the scope until the marks on the fork arm are aligned. Be as precise as possible and then press **Enter**.

At this point, on the GPS models, you will find that a GPS link may be established, providing location, date, and time information. On all other models you must provide this information. Be aware that when you enter the date, you must use mm/dd/yy (US format); the most common error in Quick Alignment is using an incorrect date format. Correct entries can be confirmed, rather than retyped, by simply pressing the **Enter** button. Use the **Undo** button to correct mistaken key presses. It is not necessary to be extremely accurate in providing location and time information, but since we are not able to center two alignment stars, the accuracy of the information you provide will determine the accuracy of tracking. After providing this information the hand control will report successful alignment and will be tracking the sky at sidereal rate. If your objective is solar observing, change the tracking rate to solar.

EQ North/South Alignment

(NexStar 8/11 GPS, NexStar 5i/8i)

EQ North/South Alignment is used when the telescope is mounted on an equatorial wedge. EQ North is used in the Northern Hemisphere and, naturally, EQ South is used in the Southern Hemisphere. Note that wedge mounting your scope is mainly used for long-exposure astrophotography. For visual use, alt–az mounting is more convenient and provides more comfortable eyepiece location.

To perform an EQ North/South Alignment, set up the tripod and roughly level it. The closer you are to leveling the tripod, the closer the scope will be to locating the alignment stars. The wedge should then be attached to the tripod and the scope to the wedge. Insert a high-power eyepiece, turn on the scope, and move the optical tube perpendicular to the fork arm (the azimuth/RA axis). Select **EQ North Align** if you are in the Northern Hemisphere or **EQ South Align** if you are in the Southern Hemisphere.

If you are using a NexStar 8/11 GPS with hand control version 1.6, the scope will automatically move to "level" or, actually, perpendicular to the fork arm. If you are using a NexStar 8/11 GPS with hand control version 2.2, or a NexStar 5i/8i, verify that the marks on the fork arm are accurately aligned when prompted.

When prompted to find the meridian, point the scope directly south if in the Northern Hemisphere or directly north if in the Southern Hemisphere. The scope will still be pointed upwards due to its position perpendicular to the fork arm and the tilt of the wedge. With the NexStar 8/11 GPS, you may release the azimuth (RA) clutch to point the scope, with the NexStar 5i/8i, use the left/right arrow buttons. Be certain not to move the scope in altitude (Dec) or the procedure will fail.

Next you will be prompted to select either an **EQ AutoAlign** or an **EQ TwoStarAlign**. The **EQ AutoAlign** automatically points the scope at two alignment stars, while **EQ TwoStarAlign** will require you to find and point out two stars. These methods are similar to the standard Auto Alignment and Two Star Alignment. I will describe **EQ AutoAlign**.

After selecting **EQ AutoAlign**, if you have the GPS module, the scope will attempt a GPS link. The top line of the display will cycle through the messages "GPS Searching", "Enter if OK", and "UNDO to edit". The GPS receiver will shortly lock onto three GPS satellites and update the date, time, and location. If large buildings, hills, or other obstructions block any of your horizons, the GPS link could take some time. Also, if you have transported the scope more than a few hundred miles, additional time will be necessary to download database information from the satellites. In any case, if you become impatient waiting for a GPS link, you can use the **Up** and **Down** buttons (**6** and **9** on the keypad) to scroll through the date, time, and location settings. If they are all correct (not very likely), press **Enter** to accept and proceed with the alignment. If they need to be adjusted, press **Undo** and make any necessary corrections.

If you have a NexStar 5i/8i without the GPS module, you must provide location, date, and time information. Be aware that when you enter the date, you must use mm/dd/yy (US format); the most common error in EQ AutoAlign is using an incorrect date format. Correct entries can be confirmed, rather than retyped, by simply pressing the **Enter** button.

The telescope will now slew (point) towards the first alignment star. If you know that the first star is behind an obstruction, or if you do not feel it is a good choice, press any of the arrow buttons to cancel the slew, then press the **Undo** button to go to the next potential alignment star. After the scope has slewed to the general location of the first star, follow the directions on the hand control to center the star in the finder and press **Enter**. Your goal at this point is simply to get close enough to the star so that it is visible in the eyepiece. After pressing **Enter**, follow the directions on the hand control to center the star in the eyepiece and press the **Align** button. Repeat this process for the second alignment star. If you centered the correct stars, the hand control will report success and the scope will be tracking in EQ North or EQ South mode. If you did not center the correct stars, you will be informed that the alignment failed and you will need to start over from the beginning.

Note that it is common for the scope to report a slew limit warning during EQ North/South Alignment. Slew limits on a wedge are still in reference to the base of the scope. This helps to protect equipment such as cameras, commonly used when equatorially mounted. Press **Enter** to override the slew limit and then keep a close watch on your equipment. If it appears an accessory will make contact with the base, press any of the arrow buttons to cancel the slew. Then press **Undo** to continue to the next potential alignment star. After completing the EQ AutoAlign method, Filter Limits will be correct for the local horizon rather than the base of the scope. To correctly apply Filter Limits when using EQ TwoStarAlign, press the **Menu** button and select **View Time-Site** to attempt a GPS link, or use **Scope Setup**, **Setup Time-Site** to manually update the date, time, and location information.

Wedge Alignment

(NexStar 8/11 GPS, NexStar 5i/8i)

This is not truly an alignment method, in the sense that it does not result in the telescope building a model of the sky. Rather, it is a utility that simplifies polar alignment of the wedge. In other words, it assists in aligning the RA axis with the Earth's axis of rotation. Wedge Alignment differs slightly depending upon the version of hand control, so it is necessary to describe each version separately

Hand Control Version 1.6 Version 1.6 is only used on NexStar 8/11 GPS scopes. The procedure is quite simple. After powering the scope, press the **Menu** button, use the **Up** and **Down** buttons (**6** and **9** on the keypad) to select **Utilities** and then **Wedge Align**. Next, select **North** for Northern Hemisphere or **South** for Southern Hemisphere. You will then be prompted to press **Enter** to find level. After you press **Enter** the scope will move the optical tube perpendicular to the fork arms. You might get to read part of a message about improving accuracy by using the **Calibrate Level** utility, described earlier in this chapter.

You will next need to confirm your date/time/location are correct. Press the **Up** and **Down** buttons (6 and 9 on the keypad) to scroll through the information. If all items are correct, press **Enter** to continue; if corrections are necessary, press **Undo** and make the required entries. Alternately, *before* you start the **Wedge Align**, you can use **View Time-Site** on the **Menu** to get a GPS link for this information.

When prompted to find the meridian, point the scope directly south if in the Northern Hemisphere or directly north if in the Southern Hemisphere. The scope will still be pointed upwards due to its position perpendicular to the fork arm and the tilt of the wedge. You may release the azimuth (RA) clutch to point the scope or use the left/right arrow buttons to slew with the motors. Be certain not to move the scope in altitude (Dec) or the procedure will fail.

The scope will now slew to a single alignment star in order to build an approximate model of the sky. As prompted on the hand control, center the star in the finder and then the eyepiece. The scope will now slew to Polaris (Sigma Octanis in the Southern Hemisphere). Adjust the wedge and/or tripod (don't use the arrow buttons on the hand control!) to center Polaris in the finder and then the eyepiece. Your wedge is now fairly well aligned. Release the altitude clutch and move the scope roughly perpendicular to the fork arms and re-engage the clutch. Power off the scope, power back on and perform an EQ North or EQ South Alignment. You are now ready to go.

Hand Control Version 2.2 or Higher

In hand control version 2.2 or higher, the wedge alignment adds more precision, requiring a full two star alignment prior to aligning the wedge. This does add some steps to the process, but should result in a more accurate wedge alignment.

Start by executing the EQ North Alignment or EQ South Alignment as described previously in this chapter. Then press the **Menu** button, use the **Up** and **Down** buttons (6 and 9 on the keypad) to select **Utilities** and then **Wedge Align**. Then select **North** for Northern Hemisphere or **South** for Southern Hemisphere. You will then be warned that continuing will lose your current alignment; naturally you will press **Enter** to continue with the Wedge Align.

The scope will now slew to Polaris (Sigma Octanis in the Southern Hemisphere). Adjust the wedge and/or tripod (don't use the arrow buttons on the hand control!) to center Polaris in the finder and then the eyepiece. Your wedge is now fairly well aligned. With the NexStar 5i/8i, use the arrow buttons on the keypad to move the scope perpendicular to the fork arm, aligning the marks on the fork arm. With the NexStar 8/11 GPS, release the altitude clutch and move the scope roughly perpendicular to the fork arms, and then reengage the clutch. Power off the scope, power back on and perform an EQ North or EQ South Alignment. You are now ready to go.

Wedge Alignment and EQ North/South Alignment on Older Models

(NexStar 5/8, New GT, Original GT)

Wedge Alignment and EQ North/South Alignment can be performed on the other NexStar models with a little ingenuity. First, we need to make a mark on the fork arms to indicate the location when the tube is parallel to the fork arm. While indoors, power your scope and move the optical tube to the marks you created for the perpendicular position. Power off and then power back on and press the Menu button. Scroll down to the **Get Alt-az** option and press **Enter**. Then, using the up/down arrow buttons move the scope up until the display indicates an altitude of exactly 90°. Note that moving past 90 does not increase the number, so you must be careful to stop right at 90. At this point, the optical tube is pointed straight up, parallel to the azimuth axis. Make another alignment mark on the tube holder for use in the field.

When in the field, move the scope to the 90° mark and then adjust the wedge and tripod (don't use the arrow buttons on the hand control!) to center Polaris in the finder, then the eyepiece. Your wedge is now fairly well aligned. For more precise alignment, required if you will be taking long-exposure astrophotographs, you must determine how far and in what direction Polaris should be offset from the center of the eyepiece. A similar technique could be used to offset the optical tube and adjust the wedge to point at Sigma Octanis in the Southern Hemisphere.

You must now select **Two Star Align** on the hand control (for the original GT, you must use Auto Align and set your location as your normal longitude, but 89° north latitude). When prompted to point level or horizontal, move the scope down to your usual perpendicular alignment marks on the fork arm. Continue with a normal alignment as described earlier in this chapter. After the alignment is complete, press the **Menu** button and set the **Tracking Mode** to **EQ North** (Northern Hemisphere) or **EQ South** (Southern Hemisphere).

Note that when the scope is mounted on a wedge, slew limits (for the new GT) and horizon limits (for the NexStar 5/8) are still in reference to the base of the scope. This can help protect equipment that might make contact with the base, such as cameras. If this is not an issue with your equipment, the new GT allows adjustment of the slew limits as described in Chapter 5, or you can override the warning each time by pressing **Enter**. There is no similar adjustment on the NexStar 5/8, so you must manually override the horizon warning by pressing **Enter**.

Filter limits on the GT models (new GT only) will be in relation to the base of the scope. Thus the many objects that are outside of your filter limits will not show on the hand control as they are below the base of the scope. To combat this, set the filter limits to +90 and –90 when using the new GT scopes on a wedge and use care when slewing to objects.

Re-Alignment

(All models)

Re-Alignment, called Third Star Alignment on the original NexStar 5 and 8, is used to replace one of the two existing alignment points (stars) with a new alignment point. This can be useful during a long observation session. Several hours after your initial alignment you might find that GoTo accuracy is starting to suffer, particularly in the eastern sky. Realigning on an object in the east will improve accuracy, especially for that section of sky. This is also useful when imaging objects too faint to see in the eyepiece. Realign on a nearby visible object, then GoTo the faint object.

Also, Re-Alignment can be useful for daytime observing. Generally we use Quick Alignment or Auto Alignment for daytime observations, but since we do not center two alignment stars, GoTo is in no way accurate. If you can locate two bright objects, say the Moon and Venus, you can realign on them and then enjoy fairly accurate GoTo performance.

Steps for Re-Alignment

- First GoTo the object you want to use as your new alignment point. For example, you could press the **List** button, then scroll to **Named Stars,** and then select a star in the eastern sky. You must then press **Enter** to GoTo the star.
- After you get the star in the eyepiece, press the Undo button enough times to return to the **NexStar Ready** or **NexStar GPS** display. (**Note:** With the original GT hand control on little NexStars, do not press **Undo,** but rather press the **Up** button then **Down** button to reselect the object in the display.)
- Press the **Align** button to initiate the Re-Alignment.
- The display will then prompt you to choose which existing alignment star you wish to replace. Use the **Up** and **Down** buttons (**6** and **9** on the keypad) to select the correct star.
- Check to be sure the new star is centered in the eyepiece and press the **Align** button (with the original GT hand control, press **Enter**) to complete the Re-Alignment.

There are a couple of things to keep in mind when using Re-Alignment. First, be sure when selecting replacements that you follow our previous rules about star selection as stated at the beginning of this chapter. Second, while it is possible to use any object from the hand control database as an alignment point, some objects are better candidates than others. For example, when you have a large, diffuse nebula in the eyepiece, how will you know it is centered? Stars are obviously the best choices, although compact deep sky objects such as small galaxies and planetary nebulae can be good choices. Due to the complex calculations involved in determining their locations, I would only recommend the Moon and planets for daytime realignment.

With hand control version 2.2 or higher on the NexStar 8/11 GPS and the NexStar 5i/8i, you can also use the Sun as an object for realignment. This permits an accurate alignment during the day allowing you to accurately GoTo planets and bright stars. For this procedure you **MUST** have a safe solar filter (see Chapter 8), as you will replace the two alignment stars with the Sun. You cannot be too careful when using your telescope in the daylight; even the briefest glimpse of the unfiltered Sun through a telescope will cause permanent eye damage or even blindness.

First, perform a Quick Alignment. Then, with the solar filter mounted on the scope (if you have a finder scope be sure it is capped to prevent lighting your hair on fire!), center the Sun in the eyepiece. Press the Align button on the hand control and select the first alignment star for replacement. Press the zero (**0**) button and follow the directions to center the Sun and press **Align.**

Next you will want to replace the second alignment star. You can either try to GoTo and locate a bright planet or star, and then use the Re-Alignment as described earlier in this section, or you can wait 10 or more minutes, recenter the Sun and repeat the steps in the last paragraph.

Finally, I would also note that I rarely use Re-Alignment as my scopes put objects in the field of view for as long as I care to stay out observing. A little care in your initial alignment goes a long way. One time you will find Re-Alignment very useful is when imaging an object too faint to see in the eyepiece, a technique discussed in Chapter 11.

Alignment Myths

Due largely to the brevity and simple approach of the user manual, several "theories" have surfaced intending to improve accuracy in the alignment process. Some of them seem sound, but due to the actual design of the programming in the NexStar, they just don't work out. I will conclude this chapter with a list of the most common alignment myths and the reasoning to dispel them.

Myth 1 – You must work as fast as you can between centering the first and second alignment star

Sounds good, just isn't true. In a test of this statement, I started an alignment, centered the first star, and then waited one hour before centering the second star. This caused no problems; GoTo and tracking behaved normally. Obviously the hand control keeps track of the amount of time between centering one star and the next and incorporates this factor into its calculations.

Myth 2 – You must accurately level the tripod

It is true that a carefully leveled tripod is necessary if you want GPS or Auto Alignment to accurately point at the two alignment stars. It is also true that an unleveled tripod will slightly affect the Filter and Slew Limits by tilting their frame of reference. But other than that, leveling the tripod has no effect on GoTo accuracy. To test this, I set up my N80 with the tripod about 15° off level. I then performed an Auto Alignment, being careful to center the correct stars. After the alignment completed, GoTo and tracking performed normally. Consider the fact that no matter what angle you tilt the tripod to, there is somewhere on Earth where it would be level.

Myth 3 – Longitude and latitude must be precise for Auto Alignment to provide good GoTo accuracy

Longitude and latitude are used during Auto Alignment simply to point near two bright alignment stars. The actual model of the sky, used for all future GoTo operations, is created based solely upon the locations recorded for the two stars. Those locations are based upon the moment when you press the **Align** button, not upon the longitude and latitude you entered earlier. Your longitude and latitude need to be accurate enough so that you are not confused as to which stars the telescope wants you to center.

Myth 4 – You must point accurately at north during Auto Alignment

Just as with longitude and latitude, pointing north is only used to point near the two alignment stars. As long as you are able to determine which stars you should be centering in the finder and eyepiece, you started close enough to north.

There is one exception to myths 2, 3, and 4. When performing an alignment during the day, accuracy when leveling the tripod, providing longitude and latitude, and when pointing out **true** north is critical for improving GoTo and tracking when you will not be able to see and center the alignment stars.

Myth 5 – For the original NexStar 5/8 and the NexStar 60/80/114/4, starting with the optical tube precisely perpendicular to the azimuth axis is not that important

A good number of people, after hearing the importance of starting alignment with the tube perpendicular to the azimuth axis, go their own way, formulating other methods for improving accuracy. Occasionally one of them will write me, declaring perpendicular is not the key; they have found a better method that is just as accurate. When asked how they improved their accuracy, the answer always turns out to be accurate leveling of the tripod, followed by accurate leveling of the optical tube, a process that takes them at least five or more minutes each night. Of course, what they are accomplishing is setting the optical tube perpendicular to the azimuth axis, but wasting five minutes of observing time to do it!

Additionally, it is mathematically possible to accurately model the sky based on two stars, without assuming all three axes (optical, azimuth, altitude) are orthogonal at the beginning of the process. But tests show that this method was not used on the older NexStars. (The newer models – the NexStar 8/11 GPS and NexStar 5i/8i – do use a method not sensitive to an orthogonal start position.) The only variable in the orthogonal condition is the angle of the optical and azimuth axes – in other words, our part in insuring an orthogonal startup is to set the optical axis perpendicular to the azimuth axis.

So, how important is it to start perpendicular? Try this. Set up a very level tripod. Move the optical tube to level, *then point it upward about 5°*. Perform an alignment, being careful to center the correct stars. Check your GoTo accuracy. I'm certain the results will convince you to adopt alignment marks on the fork arm of your NexStar 5/8 or NexStar 60/80/114/4 to indicate the position of perpendicular.

Chapter 5

Basic Operation

The NexStar telescopes are the easiest-to-use GoTo system on the market. Within a few nights you will hardly notice you are pushing buttons at all. Following are descriptions of the features you will find in the computerized hand control; I will not discuss the manual hand control available on some models.

The designations in parentheses indicate which models include each feature. For information about various versions of hand controls, refer to Chapter 3. Button and menu names are shown in **bold** print.

GoTo an Object

(All models)

The simplicity of using a NexStar is most apparent when accessing many of the catalogs (lists of objects) stored in the hand control. The Messier, Caldwell, NGC, and SAO Star catalogs are each accessed via a dedicated button on the number pad. The Moon and planets are accessed similarly via the **Planet** button. To slew to (GoTo) an object in any of these lists, simply press the correct button, key in the catalog number of the object or select it from a presented list, and press **Enter** – a simple procedure that anyone can operate after a quick orientation.

Keep in mind that on the newer models, if you request a GoTo to an object outside of your slew limits (described later in this chapter) you will receive a warning on the display. If you think the scope can safely slew to the object, you can override the warning by pressing **Enter**, otherwise, press **Undo** to cancel. Similar warnings appear on the original NexStar 5 and 8 if a requested object is below the horizon.

Note that prior to pressing **Enter** to start a GoTo, you can press the **Info** button to display information on an object. Press the **Up** and **Down** buttons (**6** and **9** on the keypad) to scroll through the information available. This is particularly useful for deep sky objects such as those in the Messier and NGC catalogs. The information presented will tell you the type of object (open cluster, galaxy, etc.) and magnitude of the object. This might prompt you to bypass attempting to spot a magnitude 13 galaxy in your NexStar 80, as it would not likely be visible. Also, since the original GT control on the

Manual Slew Rate

When using the four arrow buttons to manually point the telescope you can vary the slew speed to match your objectives. Rates 5 and 6 are generally used when centering objects in a low- or medium-magnification view, while rates of 3 and 4 are more useful at high magnification. Rates 7 through 9 are mostly used when slewing long distances or when centering objects in the finder. Rates 1 and 2 are used for manual guiding of long-exposure astrophotographs and move at very slow rates.

To change the rate at any time, regardless of the current operation pending on the hand control, press the Rate button followed by the number of the desired rate.

The arrow buttons also have a "double-button" feature that allows you to slew at rate 9 without changing the current rate. Press and hold the arrow button to start moving in the direction desired, then press and hold the opposite button to jump to rate 9.

little NexStars does not have filter or horizon limits, you might find the Altitude entry in Info to be useful to prevent those embarrassing attempts to slew towards an object under the ground!

When viewing **Info**, pressing 1 on the keypad speeds up the scrolling text. Each time you press it, the faster the scrolling becomes. Pressing 4 on the keypad slows down the scrolling; again, each time you press 4, the slower the display scrolls.

Pressing 7 on the keypad "pauses" the scrolling text. Pressing 7 again resumes scrolling.

To cancel a slew, press any of the arrow buttons on the hand control. This is useful if you notice that the scope will make contact with the mount or determine that you selected the wrong object.

Due to the different sizes of these lists, Celestron has provided slightly different access methods for the smaller lists and the larger lists. Following are descriptions of the various catalogs and, when required, examples of their use.

The Moon and Planets

Press the **Planet** button, use the **Up** and **Down** buttons (6 and 9 on the keypad) to scroll through the selections, and then press **Enter** to slew to the displayed object. If you are using one of the newer models with Filter Limits (described later in this chapter), objects outside of your limits will not be displayed even though they are visible in the sky. To override this, adjust your Filter Limits. To present a list of all planets and the Moon regardless their current location, set the Filter Limits to –90 and +90.

Messier and Caldwell Catalogs

The Messier catalog is considered one of the premier lists of deep sky objects. It offers a variety of DSOs from open clusters to galaxies. The Caldwell catalog is a good supplement to the Messier list; outstanding objects overlooked by the Messier list are included. Also, the Messier list is limited to objects visible to observers in the Northern Hemisphere while the Caldwell list includes objects visible only in the Southern Hemisphere.

To direct the scope to the Orion Nebula, M42, press the **M** button, key in 042, and then press **Enter**. The Caldwell catalog is accessed similarly via the **Cald** button. Note that for Messier and Caldwell objects it is necessary to enter 3-digit numbers. Use the **Undo** button to erase mistaken key presses prior to pressing **Enter**.

NGC Catalog

The NGC (New General Catalog) is a comprehensive list of more than 7000 deep sky objects, to include most of the brighter DSOs in the sky. The NGC was compiled well before the age of computerized databases, so not surprisingly there are some erroneous entries. Nonetheless, there are enough things to see to keep you busy for the rest of your observing career.

To access the NGC catalog, press the **NGC** button and key in the NGC catalog number in 4-digit fashion. For example, to slew to the Double Cluster in Perseus, NGC869 (and NGC884), press the **NGC** button, key in 0869, and then press **Enter**.

The little NexStars do not contain the entire NGC listing. The original hand control on the little NexStars simply ignores your request if the NGC object you enter isn't in the database. On the new hand control, behavior gets a little confusing. If the NGC object you enter isn't in the database, the hand control chooses the next object that *is* in the database. Generally this is not a good thing; so unless you are certain a particular NGC object is available, press the **Info** button to see what will be selected. If it is your original target, then great – simply press **Enter**. If not, maybe you will get lucky and find a new favorite.

SAO Stars

There are actually four different star lists in the NexStar hand control. The main star list is a selection of stars from the Smithsonian Astrophysical Observatory (SAO) catalog. This list is accessed via the **Star** button on the keypad. Use of the **Star** button varies according to model.

The NexStar 5/8 and the original GT hand control for the little NexStars store their star lists using a serialized number unique to these scopes. To slew to an SAO star, you must know the correct NexStar number. Cross-referenced lists are available at the Celestron web site (http://www.celestron.com), but I have created enhanced versions, including expanded double star lists. These enhanced versions are available in the Downloads section of my NexStar Resource Site – http://www.NexStarSite.com. Once you know the NexStar number for the star, press the **Star** button, key in the number, then press **Enter**.

On the new GT hand control for the little NexStars, you can GoTo stars using the SAO catalog number directly. Simply press the **Star** button, key in the SAO catalog number, then press **Enter**. Note that you must use a full six digits for the SAO number. The scope will not start the slew immediately, but rather it will display the database information on the star. Press **Enter** again to start the slew. If the star is not in the database, the next star in the catalog will be displayed instead. Press **Enter** to GoTo that star or **Undo** to cancel.

The NexStar 8/11 GPS and 5i/8i behave differently to allow the Filter Limits to be applied to the SAO list. The stars are accessed by SAO number. Press the **Star** button, key in the first four digits from the SAO number and you will be presented with a list of stars in the database that are within your current filter limits. Press the **Down** button (**9** on the keypad) to scroll through the list to the star you are looking for. If you scroll past your target, you must press **Undo** until you are back at the **NexStar GPS** prompt, press another catalog key, then **Undo** and start over by pressing the **Star** button. Look for this to be improved in future versions.

Named, Double, and Variable Stars

The named, double, and variable star lists are generally more useful to the observer than the SAO star catalog. In particular I enjoy the various double stars Celestron has made available. Appendix B provides complete lists of the stars included in these categories. For additional double stars that can be accessed by SAO or NexStar number, check the Downloads section of my NexStar Resource Site – http://www.NexStarSite.com. The **Named Star**, **Double Star**, and **Variable Star** lists can be found on the **List** button. Note that you will only be presented with objects within your Filter Limits.

The Tour List

The tour is a sampling of some of the best deep sky objects currently visible. While it won't keep you happy night after night, since each night will be largely a repeat of the previous, it is useful if you haven't prepared a list of objects you wish to see. One potential problem, if you own one of the smaller NexStar models, is that the tour does not take into account the aperture of the scope. The objects presented are basically the same whether you are using the NexStar 60 or the NexStar 11. Many of the fainter objects will not be visible in smaller scopes and many of the asterisms will be too wide to enjoy in the larger scopes. Thus you will just move on to the next object on the tour.

Press the **Tour** button on the keypad and you will be presented with the first object on the tour list that falls within your current Filter Limits. You can press **Enter** to slew to that object or the **Info** button for the object's description, magnitude, and more. When you are ready for the next tour object, use the **Up** and **Down** buttons (6 and 9 on the keypad) to scroll through the list.

Named Objects

The Named Object list is a compilation of the best deep sky objects that aren't on one of the star or asterism lists. Pressing the **Info** button for any of these objects will display their Messier or NGC designation. For a complete list of the Named Objects, refer to Appendix B. **Named Objects** is found on the **List** button. Note that you will only be presented with the objects within your current Filter Limits.

Asterisms

One of most unique lists available in the NexStar hand control is the list of asterisms. Asterisms are simply recognizable patterns randomly created by groups of stars. The name of the asterism often describes the pattern, such as "Coathanger Cluster" or "Number 7 Cluster". Most of asterisms require low magnification and a wide field of view to take in the entire shape. They are especially good targets for the NexStar 60 and 80. The Asterisms list is found on the **List** button and will be limited to objects within your current Filter Limits. Refer to Appendix B for a complete list of NexStar Asterisms.

CCD Objects

The NexStar 8/11 GPS and 5i/8i include a list of objects well suited to CCD imaging. These are generally small pairs, trios, or clusters of galaxies that will fit within a single CCD framed image. The CCD Objects list is found on the **List** button and is limited to objects within your current Filter Limits. Appendix B presents a complete list of all the CCD Objects.

IC Catalog

The IC (Index Catalog) is not as commonly used as the NGC catalog, though J.L.E. Dreyer authored both. The IC catalog is comprised of more than five thousand deep sky objects; most of them are quite faint and require a large telescope and dark skies. The IC catalog is included on the NexStar 8/11 GPS and 5i/8i, is found on the **List** button, and functions the same as the NGC list.

Abell Catalog

The Abell catalog is a compilation of galaxy clusters and the first 2712 objects from this catalog are available on the NexStar 8/11 GPS and 5i/8i. In many cases a single Abell object is a cluster of more than 50 galaxies in a single field of view! Most of these galaxies are much too faint to be seen in even the largest NexStar telescopes, but the possibilities for long-exposure astrophotography are exciting. You will find the Abell catalog on the **List** button.

GoTo a Specific Right Ascension and Declination

When the object you are interested in is not to be found in one of the included catalogs, or you just can't find which catalog it is in, all is not lost. Providing you know the RA and Dec of the object, simply press the **Menu** button, scroll to **GoTo RA–Dec**, and directly enter the coordinates.

Get the Current Right Ascension and Declination

While not a GoTo operation, this function is useful after you have located an object manually in the eyepiece. Press the **Menu** button, scroll to **Get RA–Dec**, and press **Enter**. You can then research the given coordinates and determine the object in the field of view.

Required Hand Control Setup

Some of the utilities and functions of the computerized hand control are central to maximizing the operation of your scope. A few moments spent tweaking these settings will result in a smoothly operating scope.

Anti-Backlash

(All models)

All gears, including the ones used to move a NexStar telescope, have a characteristic known as backlash. This is simply the play in the gears that shows itself as a delay when using the arrow buttons to manual slew the scope. The NexStar 8 GPS and 11 GPS use worm-driven gears, which inherently exhibit minimal backlash. All other models use spur gears, which have a good deal of backlash.

The delay caused by backlash is noticeable when using the arrow buttons to move in the direction opposite of tracking. For example, if you are in the Northern Hemisphere and viewing an object towards the south, pressing the left arrow button will require compensation for backlash, otherwise you will notice a delay before the scope actually starts moving. The NexStar scopes employ backlash compensation, or Anti-Backlash as Celestron calls it, to rewind the motor and take up the slack so that the scope starts moving immediately after pressing an arrow button. After you release the arrow button, compensation winds the motor back in the other direction to re-engage the gears and resume tracking. Since no two sets of gears are exactly the same, we can fine-tune the settings for the altitude and azimuth axis on each scope.

NexStar 5/8 and Original GT On the original NexStar 5/8 and the original GT control for the little NexStars, there is just a single setting for each axis. In other words, you can specify a value for azimuth compensation and a value for altitude compensation.

To correctly set backlash compensation, start by aligning the scope. Next, press the **Menu** button, scroll to **Anti-Backlash,** and press **Enter**. Set the **Azm Backlash** to 0 and the **Alt Backlash** to 0. **Azm Backlash** corresponds to the left/right buttons, while **Alt Backlash** corresponds to the up/down button. While viewing an object in the eyepiece, observe the responsiveness of each of the four arrow buttons. Note which directions exhibit a pause, for example, left and down. Working one axis at a time, adjust the backlash setting high enough to cause immediate movement without resulting in a pronounced jump when pressing or releasing the button.

NexStar 8/11 GPS, NexStar 5i/8i, New GT On the newer models, there are two values for each axis, positive and negative. Positive is the amount applied when you press the button, rewinding to get things moving quickly, while negative is the amount applied when you release the button, winding back the other direction to resume tracking. Often they should be set the same; however if you observe smooth motion when you push the button, but the object jumps when you release the button, lower the negative value.

To correctly set backlash compensation, start by aligning the scope. Next, press the **Menu** button, and select **Utilities** on the little NexStars or **Scope Setup** on the bigger models. Scroll to **Anti-Backlash** and press **Enter**. Set the **Azm Positive** and **Negative** to 0 and the **Alt Positive** and **Negative** to 0. **Azm** values correspond to the left/right buttons, while **Alt** values correspond to the up/down button. While viewing an object in the eyepiece, observe the responsiveness of each of the four arrow buttons. Note which directions exhibit a pause, for example, left and down. Working one axis at a time, adjust both the positive and negative backlash settings high enough to cause immediate movement without resulting in a pronounced jump when pressing or releasing the button. At this phase, use the same values for positive and negative. If you note a jump when releasing the button, but find that setting both values lower results in a pause when pressing the button, go with the higher value for positive, but use a lower value for negative.

Note that on the new GT there is a slight pause after you release a button until the backlash compensation is applied. If you attempt to use the opposite button before compensation is complete, the scope will not respond. A little patience goes a long way towards smooth operation in this case.

Slew Limits

(NexStar 8/11 GPS, NexStar 5i/8i, New GT)

The Slew Limits feature is designed to prevent the telescope, or equipment attached to it, from making contact with the tripod or base. On the NexStar 60/80/114, the optical tube can easily make contact with the tripod when pointing at objects high in altitude. With other models, adding a camera or longer optical accessories to the back of the scope might result in a similar situation. Slew Limits control the minimum and maximum altitudes (angles from the horizon) that a slew can attempt without displaying a warning. To continue to protect your equipment even when mounted on a wedge, Slew Limits are measured in relation to the base of the telescope, rather than your local horizon.

Keeping in mind that 0° is the horizon and 90° is straight up, the following procedure will accurately set the maximum Slew Limits required for your equipment:

- Set up the tripod, point north and line up your perpendicular marks or level everything if you haven't adopted the marks.
- Do a Quick Alignment.
- Manually slew to the point you feel is as high as you can safely go.
- Use the menu option for Get Alt-az.

The altitude displayed is your maximum setting for slew limits. For minimum, I recommend leaving the default setting of 0° in place.

To access Slew Limits, press the **Menu** button, then select **Utilities** on the little NexStars or **Scope Setup** on the bigger models. Scroll to **Slew Limits**, press **Enter** and you will be presented with selections for minimum and maximum limits. Any future attempts to GoTo an object outside of your Slew Limits will result in a warning that can be ignored by pressing **Enter**, but if you do, closely monitor the telescope to prevent contact with the tripod or the base. If there is contact, your alignment will be lost and you will need to power off the scope and start again with a new alignment. Additionally, the motors in the larger NexStar models are strong enough to damage equipment if contact is not prevented.

Filter Limits

(NexStar 8/11 GPS, NexStar 5i/8i, New GT)

Filter Limits are also set as minimum and maximum altitude. On the little NexStars this altitude angle will always be in relationship to the base of the telescope, but on the bigger NexStars, Filter Limits are in relation to your actual local horizon. Objects outside of your Filter Limits will not be displayed as you access the various lists in the hand control. For example, if your minimum filter limit is set to 30°, but Venus is at 20°, Venus will not be displayed as an available object on the **Planet** button.

I generally leave the Filter Limits set to 0° and 90° since my horizon is fairly clear in most directions. With the Filter Limits set to 0° and 90°, the objects in the list include anything currently above the horizon. Slew Limits will still be in effect to protect your equipment and alignment. As discussed before, pressing **Enter** will override the slew limit warning if necessary.

On the little NexStars you will find **Filter Limits** directly on the **Menu** button. For the NexStar 8/11 GPS and 5i/8i, press the **Menu** button then select **Scope Setup**. Scroll to **Filter Limits**, press **Enter** and you will be presented with settings for minimum and maximum limits.

On the NexStar 8/11 GPS and NexStar 5i/8i, the Filter Limits are in respect of your local horizon, rather than the base of the scope, useful when mounted on a wedge. If you perform an EQ Auto Align, the scope has all the information it needs – date/time/location – in order to make this work. But, when you perform the EQ Two Star Align, you must provide the required information. If you have the NexStar 8/11 GPS or the NexStar 5i/8i with the optional GPS module, select **View Time-Site** on the **Menu** button to get the information from the GPS system. If you don't have GPS, use **Setup Time-Site** on the **Scope Setup** menu to manually enter the date, time, and location.

Compass and Level (Altitude Sensor) Utilities

(NexStar 8/11 GPS, NexStar 5i/8i with optional GPS module)

The internal compass and downstop switch ("level" or "altitude" sensor) on the NexStar 8/11 GPS and the NexStar 5i/8i with optional GPS module are roughly calibrated at the factory. Both should be calibrated for improved accuracy. Calibrating the compass is necessary to compensate for the difference between magnetic and true north at your location, allowing the scope to come much closer when initially pointing at the two alignment stars. Calibrating level is required to determine the point where the optical tube is perpendicular to the fork arms, necessary to provide accurate GPS Alignments as well as accurate EQ North/South and Wedge Alignments on the NexStar 8/11 GPS.

- Start by very carefully leveling the tripod in all directions prior to attaching the scope. Set the level on the head of the tripod, testing alternately across the top of two legs at a time. Adjust the legs until level in all orientations.

- Next perform a GPS alignment. Take care to accurately center the alignment stars and, after completion, check to be sure that GoTo is accurately placing objects in the field of view.

- Press the **Menu** button, select **Utilities**, **Compass**, and then **Calibrate**. Hand control version 1.6 will not give any indication that the calibration was successful, but in fact the calibration will have been completed after pressing Enter. Version 2.2 or higher reports success.

- On the **Menu** button, select **Utilities**, **Level** (or **Alt Sensor**), and then **Calibrate**.

Note that even after calibrating the compass and level, during alignment, the scope will still point off from true north and a true perpendicular position. The calibration routines do not change the positioning of the internal compass or downstop switch; rather they store compensation values used in future alignments.

If you travel a significant distance with your scope, the new location may have a different magnetic north deviation and thus require the compass to be recalibrated. You should not need to recalibrate level unless you reset the hand control to factory defaults or return your hand control to Celestron for servicing.

RS-232 Mode

(NexStar 5/8, New GT)

The RS-232 mode is found on the **Menu** of the original NexStar 5/8 and the new GT hand control for the little NexStars. For these models, RS-232 mode is necessary for communicating with an external computer connected via the RS-232 port on the bottom of the hand control. This is discussed in detail in Chapter 10. After successful alignment, to enable RS-232 mode, press the **Menu** button, scroll to **RS-232**, and then press **Enter**. To cancel RS-232 mode and terminate communication with the external computer, press the **Undo** button.

GoTo Approach

(NexStar 8/11 GPS, NexStar 5i/8i)

GoTo Approach is discussed more fully in the next section of this chapter, but some changes are required in most situations. You will find **GoTo Approach** on the **Scope Setup** menu. Both axes, altitude and azimuth, can be set to either **Positive** or **Negative**. In altitude, **Positive** means the final approach will be with the front of the scope moving up, azimuth **Positive** indicates the front of the scope is moving to the right, both as viewed from the back of the scope.

As noted in Chapter 4, I recommend you change the altitude GoTo Approach to **Positive** on the NexStar 8/11 GPS and the NexStar 8i to take advantage of the default final movements of the GPS Alignment routine and the natural back-heavy balance of these scopes. This also matches with the final movements for centering the alignment stars as presented in Chapter 4.

If you do not change the altitude approach, you must make all final movements during alignment to move the front of the scope down, rather than up as recommended in Chapter 4.

If you are observing from the Southern Hemisphere, change the azimuth GoTo Approach to Negative for NexStar 8/11 GPS and NexStar 5i/8i telescopes. The majority of celestial objects observed by those in the Southern Hemisphere are located to the observer's north. Such objects are moving to the scope's left, as are all objects when mounted on a wedge. An azimuth GoTo Approach of Negative causes the final GoTo motion to be towards the left, starting tracking without delays due to gear backlash.

If you do not change the azimuth approach, you must make all final movements during alignment to move the front of the scope right, rather than left as recommended in Chapter 4.

Less Commonly Used Features

Several other features in the computerized hand control are less used but still significant.

View Time-Site and Setup Time-Site

(All models except Original GT)

The GPS Alignment, Auto Alignment, Quick Alignment, and EQ Auto Alignment all require time and site information to be established at the time of alignment. If you use Two Star Alignment or EQ Two Star Alignment, it is necessary to provide the scope with time information prior to directing the scope to slew to the Moon or a planet. Time information consists of the current date and time, your local time zone, and daylight savings or standard time.

Use **View Time-Site** (on the **Menu** button) to update the time and site location on any NexStar with GPS or simply to view such information on any model. On the NexStar 8/11 GPS and NexStar 5i/8i, you will also find **Local Sidereal Time** listed in the information provided. On the little NexStars, **Local Sidereal Time** is a separate selection on the **Menu** button. Local sidereal time is the line of right ascension currently crossing the meridian (straight overhead).

Use **Setup Time-Site** to manually enter time and site information. On the NexStar 8/11 GPS and 5i/8i, press the **Menu** button, then select **Scope Setup**, and then **Setup Time-Site**. On the little NexStars, press the **Menu** button, select **Utilities**, and then **Setup Time-Site**.

The original NexStar 5 and 8 do not actually have a separate view and setup function, but instead you will find **Date/Time** on the **Menu** button. This allows you to enter the current date and time, although no other site information is available.

User Defined Objects

(All models)

NexStar telescopes have the ability to store several user-defined objects for future GoTo operations. I find this feature useful for a few situations. First, you discover one of your favorite objects is not available in any of the existing lists. Or, you are manually searching an area of the sky and discover an object whose coordinates you wish to save. In these cases you will be storing a "sky object". It can also be useful during the day, if you happen to be using your NexStar as a spotting scope. In this case you will be saving a "land object".

The NexStar 8/11 GPS and NexStar 5i/8i can store 200 sky objects and 200 land objects. The little NexStars with the new GT hand control can store 25 sky objects and 25 land objects. The original NexStar 5/8 can store 20 sky objects and 5 land objects. All objects are permanently saved.

There are four operations regarding User Defined Objects:

Save Sky Object   After the object is centered in the eyepiece, press the **Menu** button, select **User Objects**, and then select **Save Sky Object**. Enter or select the number that will designate the object and it will be stored. On the original NexStar 5/8, you are limited to the numbers 1 through 20. The current right ascension and declination will be stored along with the number you have chosen. At this point it would be best to make a note to record the number you attached to the object.

Save Land Object   The scenario for land objects is more complicated. The azimuth and altitude are stored for the object. When using your NexStar as a spotting scope, you will not perform an alignment and the scope will not be tracking. Altitude is still measured as the angle in degrees above the horizon. Azimuth is a measurement of degrees, starting with 0 and continuing completely around the circle to just before 360. The 0 azimuth point is the direction the scope was pointed when power was applied. Thus to use a set of land object coordinates in the future, the scope must be oriented exactly the same each time the scope is powered up. If you are using your scope from your back patio, I would recommend fully extending the legs of the tripod and marking the location of the feet. Then sight a distant, immovable object and power off. This will be your starting point each time you use the scope.

After an object you would like to save is centered in the eyepiece, say a bird's nest in a distant tree, press the **Menu** button, select **User Objects,** and then select **Save Land Object.** Enter or select the number that will designate the object and it will be stored. On the original NexStar 5/8, you are limited to the numbers 21 through 25. The current altitude and azimuth will be stored along with the number you have chosen. At this point it would be best to make a note to record the number you attached to the object.

Enter RA & Dec Use this option to directly store the right ascension and declination of a sky object without aligning the scope or centering the object in the eyepiece. This can be useful to create a list of objects for a custom tour. Press the **Menu** button, select User Objects, and then select **Enter RA & Dec.**

GoTo User Object Press the **Menu** button and select **User Objects.** On the original NexStar 5/8 and the GT hand control for the little NexStars, you next enter the object number. On all other models you select either **GoTo Sky Obj** or **GoTo Land Obj.** On the NexStar 8/11 GPS, NexStar 5i/8i and new GT hand control, use the **Up** and **Down** buttons (**6** and **9** on the keypad) to scroll through the user objects.

Light Control

(All models except Original GT)

You might find that the hand control lighting is too bright when you are at a truly dark site. While there is no provision to adjust the brightness, there are settings to turn off the backlight. The keypad and LCD panel are controlled separately. On the original NexStar 5/8, you will find **Light Control** after pressing the **Menu** button. On the other models, press the **Menu** button, select **Utilities,** and scroll down to find **Light Control.**

Cordwrap Prevention

(NexStar 8/11 GPS, NexStar 5i/8i, NexStar 5/8)

Unlike some other computerized GoTo telescopes, the NexStar line has no mechanical stopping point in azimuth and thus is able to take the shortest route during a slew. If you have power cords, computer cables, or camera cables attached to the moving scope, that could present a problem as the scope may wrap the cords around the base. **Cordwrap** is found under **Scope Setup** on the **Menu** button on the NexStar 8/11 GPS and NexStar 5i/8i models. On these models it is on by default after an EQ Alignment, but off by default after other modes of alignment. On the NexStar 5/8 **Cordwrap** is on by default and is found by scrolling down after pressing the **Menu** button.

If I'm not imaging, I always prefer to have **Cordwrap** off and in any case it is not available on the little NexStars. If the cord does wrap around the base, there are ways to deal with it. If you are using an a.c. adapter to power your scope, try to have enough cord so that you can lift it up over the scope and unwrap it. If you are using a battery pack, pick it up and move it around the scope or if it is the small battery pack that comes with the NexStar 60/80/114, hang it on the fork arm. If all else fails, you can simply slew the scope in the reverse direction to untangle the cords.

Periodic Error Correction

(NexStar 8/11 GPS)

Although the worm-driven gears in the NexStar 8/11 GPS are very precise, they do exhibit small imperfections that are significant when the scope is used for long-exposure astrophotography. Such imperfections produce a repeating cycle of errors: periodic errors. The hand control is capable of correcting periodic errors in the right ascension axis with the Period Error Correction (PEC) feature. Correcting just the RA axis is sufficient, since the declination axis is not used for tracking when mounted on a wedge – a requirement for long-exposure astrophotographs. Note that if the periodic error is relatively small on your scope, an autoguider will likely do fine without PEC. But, if the periodic error is too great for an autoguider (I have not heard of a NexStar 8/11 GPS that had such high periodic error), or if you will be manually guiding exposures, PEC could significantly improve your photographs.

To use PEC during a photograph, you must first store the periodic error (unique to each scope) by "recording" an 8-minute session of manual guiding. The worm gear that drives the RA axis takes 8 minutes to complete one turn. This recording is permanently stored in the hand control and will not need to be remade unless you return your hand control to Celestron for servicing. Once recorded, you simply "play back" the PEC recording during an imaging session.

To make the PEC recording, set up the scope on a wedge with an accurate polar alignment. Perform an EQ Auto Alignment or EQ Two Star Alignment and insert a reticle eyepiece in the scope. Choose a fairly bright star near the celestial equator (0° Declination) as a guide star and center it in the reticle. If you are not familiar with manual guiding, spend a few moments practicing. To start the recording, press the **Menu** button, select **Utilities**, **PEC**, and **Record**. On hand control version 1.6, 5 seconds after selecting **Record**, the 8-minute recording session will begin. Note that the first time each session that PEC Record or PEC Playback is selected, the worm gear turns to its index mark for reference. This will generally move your guide star out of the center of the eyepiece, or even out of the field of view. If so, cancel the recording after the index is found, recenter the star, and start PEC Record again.

The process is slightly different on hand control version 2.2 or higher. You will be prompted to press **Enter to Initialize**, which will move the worm gear to its index mark. Use the left/right arrow buttons to recenter the guide star and then press **Enter**, as prompted on the display, to begin recording.

For 8 minutes, you must keep the guide star centered in the reticle. Note that any drift in declination should be ignored; you should only be correcting with the left and right arrows. Drift in declination is caused by an imperfect polar alignment, something you can correct with a procedure known as a drift alignment (not discussed here). After 8 minutes, the recording will automatically stop. Now when imaging in the future you can play back the recording to reduce periodic error.

Autoguide Rate

(NexStar 8/11 GPS, NexStar 5i/8i)

The Autoguide Rate controls the speed with which an autoguider can slew the scope to make corrections during a long-exposure astrophotograph. **Autoguide Rate** is found on the **Scope Setup** menu and is expressed as a percentage of sidereal rate.

Tracking Mode and Rate

(All models)

On the original NexStar 5/8 and the original GT hand control for the little NexStars, **Tracking Rate** and **Tracking Mode** are separate items found on the **Menu** button. On the other models, **Tracking** is found on the **Menu** button and **Mode** and **Rate** are items under **Tracking**.

There are four tracking modes:

- **Alt-az** – used to track the sky when mounted in altitude-azimuth mode – mounted directly on the tripod with no wedge.
- **EQ North** and **EQ South** – used to track the sky when the scope is mounted on a polar-aligned wedge in either the Northern or the Southern Hemisphere.
- **Off** – turns off all tracking and the telescope sits still. The scope still maintains its model of the sky, so after turning on tracking by selecting one of the other tracking modes you can still use GoTo and the scope will behave normally.

There are four tracking rates:

- **Sidereal** – the rate appropriate for tracking the sky's normal motion due to the continuous rotation of the Earth.
- **Solar** – the rate appropriate for tracking the Sun's motion through the sky. This is very close to sidereal rate and other than during hours of direct, continuous solar observation you will not likely notice a difference.
- **Lunar** – the rate appropriate for tracking the Moon's motion through the sky. The Moon is a fast mover, as you can easily see by noticing its location from one night to the next. If you plan on observing the Moon for several minutes, selecting lunar tracking will help to keep it centered, particularly on the more accurate tracking NexStar 8/11 GPS and NexStar 5i/8i.
- **King** – this rate corrects for the effect of refraction due to the thicker atmosphere low in the sky. This rate is available only on the original NexStar 5/8 and the original GT hand control for the little NexStars. Due to the less accurate tracking of these models, King rate may not prove very effective.

Hibernate

(NexStar 8/11 GPS, NexStar 5i/8i)

The Hibernate feature allows you to power off the scope while still maintaining your star alignment. When you "wake up" the scope, it is ready to go with no alignment required. This can be very useful for permanently mounted scopes. After you finish observing for the night, just Hibernate and power off. The next time you want to use the scope, simply power up and you will be observing in seconds. This also proves useful for daytime observing. You can accurately align your telescope at night when the stars are visible, Hibernate, and then cover the scope. The next day you can accurately GoTo planets and bright stars for an unusual view against a blue background. Or you can enjoy very accurate tracking of the Sun.

To use Hibernate, the scope should be aligned and tracking. Press the **Menu** button, select **Utilities,** and then select **Hibernate**. You will be prompted to position the scope and press **Enter**. At this prompt, use the arrow buttons to manually point the scope to a safe storage position, then press **Enter**. You may then turn off the power. While the power is off, do not manually move the scope or you will lose your alignment.

When you power up again, you will be prompted to press **Enter** to **Wake Up** the scope. Do **not** press the direction buttons before you successfully wake up the scope or you will lose your alignment. After you press **Enter**, if your scope has GPS, it will attempt a link or you may press **Undo** to set the correct date and time. If you do not have the GPS module on the NexStar 5i/8i, you must manually enter the time. Note that GoTo performance is entirely dependent upon accurate date and time after a Hibernate. Every 1 minute of error is as much as 15 arc minutes of error in the sky. If you are depending upon a watch for time, it is important to use the same watch you used during the initial alignment before Hibernate.

Direction Buttons

(NexStar 8/11 GPS, NexStar 5i/8i, New GT)

This feature allows you to exchange the directions of the up/down or left/right arrow buttons. This provides better control when using the scope with different combinations of visual and photographic accessories. For example, perhaps you have connected a small video camera to your NexStar 4's straight-through port and then connected this to a video monitor to give a group of spectators a view of the Moon. Changing the direction of the up/down buttons would allow more natural control as you are now using the scope without the diagonal. This feature can also be very useful when mounting a different optical tube on a NexStar mount as discussed in Chapter 13.

Direction Buttons is found on the **Utilities** menu on the little NexStars, or you will find it on the **Scope Setup** menu on the bigger NexStars.

GoTo Approach

(NexStar 8/11 GPS, NexStar 5i/8i)

GoTo Approach allows you to specify the direction the scope takes during final centering when it slews to an object. This allows the final motion to minimize the effects of backlash in the gears. You will find **GoTo Approach** on the **Scope Setup** menu. Both axes – altitude and azimuth – can be set to either **Positive** or **Negative**. In altitude, **Positive** means the final approach will be with the front of the scope moving up. In azimuth, **Positive** means the final approach will be with the front of the scope moving to the right. Both of these directions are as viewed from the back of the scope. The default settings are altitude **Negative** and azimuth **Positive**.

An altitude approach of **Positive** is generally required if the telescope is back-heavy. The NexStar 5i is nearly balanced and so an altitude approach of **Negative** is appropriate. The NexStar 8/11 GPS and 8i are back-heavy and thus benefit from an altitude approach of **Positive**.

It should be noted that the direction of final approach determines the directions you must move the scope when centering the alignment stars and moving to the index position. The directions given in Chapter 4 depend upon changing the altitude approach to **Positive** for the NexStar 8/11 GPS and NexStar 8i. Also, if you are observing from the Southern Hemisphere it is important to change the azimuth approach to **Negative** for the NexStar 8/11 GPS and the NexStar 5i/8i.

Get Alt–az and GoTo Alt–az

(All models)

Since these options were ostensibly designed for daytime observing of land objects, you will not likely find them useful. Occasionally I do use **Get Alt–az** to get a precise reading of the altitude of an object when the scope is alt–az-mounted. This can be a useful piece of information to enter into your observation log as it often proves instructive when trying to understand why an object was clearly observed one night (high in altitude), but barely visible another (lower in altitude).

On the little NexStars and the original NexStar 5/8, you will find **Get Alt–az** and **GoTo Alt–az** by scrolling down after pressing the **Menu** button. On the NexStar 8/11 GPS and NexStar 5i/8i, press the **Menu** button, select **Utilities** and scroll down to locate these options.

Select Model

(New GT, Original GT)

While other hand controls automatically detect the model of NexStar scope they are connected to, the hand control on the little NexStars must be told. On the new GT hand control, press the **Menu** button, select **Utilities**, and scroll to **Select Model** and press **Enter**. Scroll up or down to display the correct model and press **Enter**.

The procedure is a little more complicated on the original GT hand control. If you select the model by name, for example, NexStar 80, your scope will not work correctly. The scope will not be very close to the two alignment stars, nor will later GoTo slews be accurate. In fact, if you get the scope to say it was aligned successfully, it might miss objects by many degrees during GoTo – even when you attempt to GoTo one of the two alignment stars!

To correctly set the model on the original GT hand control, press the Menu button, scroll to **Model Select**, select **Custom**, and then enter the following numbers for both **AZM** and **ALT**:

- NexStar 60 and 80: **0726559**
- NexStar 114 and 4: **1059334**

Reset to Factory Settings

(NexStar 8/11 GPS, NexStar 5i/8i, New GT, Original GT)

This option resets many of the stored parameters in the hand control to the initial factory settings. Options such as backlash settings, date/time/location, slew and filter limits, and direction button settings will be reset. PEC recording and user-defined objects are not reset. In other words, if a stored parameter is originally blank, it is not erased by a reset. Occasionally, a misbehaving telescope is corrected by a reset to factory settings.

For the NexStar 8/11 GPS and 5i/8i, press the **Menu** button, select **Utilities**, and then select **Factory Settings**. Press **0** on the keypad to reset or **Undo** to cancel. For the GT models, use **Select Model** as described above.

Version

(NexStar 8/11 GPS, NexStar 5i/8i)

Use the Version option to display the firmware (internal program) versions for the hand control (HC) and motor control (MC). On hand control version 1.6, this menu option displays the version of the HC on the top line of the display and two numbers for the MC version on the second line of the display. The first of the two MC numbers is the firmware version for the azimuth, and the second is for the altitude. On hand control version 2.2, this menu option displays the HC and MC versions on the top line of the display and the GPS and serial bus control on the second line. On the top line, the first number is the HC version, the second is the MC azimuth version, and the third is the MC altitude version. On the second line, the first number is the GPS version and the second number is the serial bus version.

To display the version, press the **Menu** button, select **Utilities**, and then select **Version**.

Chapter 6

Expanding Your Horizons – Choosing Objects to View

After your first exciting looks at the Moon and planets, you will likely try out the objects in the Tour list. Some of those objects may not be visible with your model of NexStar scope and in any case the Tour list will only keep you occupied for so long. How do you decide what to view? Do you find yourself simply pressing the "M" button and starting at "001"? How do you go about maximizing your limited time under clear skies? There are probably as many ways to prepare for an evening as there are people with telescopes! But there are some suggestions I can give you, as well as resources and ways to use them.

Understanding What Can Be Seen in Your Scope

The faintest object you can see depends directly upon the aperture of your telescope. As discussed in Chapter 2, the light-gathering power of a telescope is a function of the area of the objective. Relatively small increases in aperture result in fairly dramatic increases in image brightness. For reference, Table 6.1 shows limiting magnitude under good, dark sky conditions. From the table, we should expect that a NexStar 4 would allow us to detect an object of about 11.7 magnitude when we are under dark skies. Unfortunately it is not quite that simple.

First, we must consider the seeing condition known as transparency. We need more than dark skies to see the faintest objects; we also need clear skies. Skies with particles of moisture, dust, or pollutants have poor transparency and make faint objects difficult to view. This is even more critical under light-polluted skies or a sky with a bright Moon. During a night with good transparency you might be able to easily see a faint globular cluster or galaxy, but the next night moisture in the air could render it invisible. Soon you will gain enough experience to judge the night sky and learn what to expect.

Second, we must recall that magnitude for deep sky objects is generally derived by taking the total luminosity of the object and reporting as if it were a single point. This is accurate for stars, as they *are* single-point light sources. For most other deep sky objects,

Table 6.1. Limiting magnitude under good, dark sky conditions

Instrument	Aperture	Limiting magnitude
Naked eye	About 7 mm	6
Binoculars	50 mm	9.5
Telescopes:		
	60 mm (2.4 in)	10.5
	80 mm (3.1 in)	11.3
	4 in.	11.7
	114 mm (4.5 in)	12
	5 in.	12.3
	8 in.	13.3
	11 in.	14
	14 in.	14.6

the magnitude reported is potentially misleading. When choosing targets for a viewing session, we must consider both the magnitude and the size of the object.

If a planetary nebula, M27 for example, is reported as magnitude 7.4 but it is 8 by 6 arc minutes in size, the result is an object with a low surface brightness. In other words, it is actually very faint. Larger galaxies and nebulae are even more misleading. M81 is magnitude 6.9, nearly naked-eye limit if it were a star. But M81 is a galaxy with a size of 20 by 10 arc minutes, resulting in a low surface brightness that makes it difficult to see without clear, dark skies.

Third, light pollution has a devastating effect on all but the brightest objects. Larger apertures can help to overcome this problem, but there is no substitute for dark skies. If you live in a small community, the background light pollution may not be too bad and you simply need to deal with a few local lights from your closest neighbors. Inviting them out to take a look through your telescope will often be enough to secure their cooperation in extinguishing outdoor lighting when you are observing. Otherwise, careful selection of where you set up can often block most lights from view.

In a city, the situation is considerably worse. The entire sky glows with the many thousands of lights flooding the city. Even if you can isolate yourself from any local lights, the skyglow will make most deep sky objects impossible to see. Planetary and lunar observations are not affected appreciably by light pollution, but for the deep sky you need to travel to a dark site. In light-polluted skies you might be able to detect a faint object, but under dark, clear skies details come into view. One of the best examples of this is the Whirlpool Galaxy, M51. Under city skies you can generally make out the core of M51 and its small companion galaxy, but nothing more. Under dark, transparent skies you can see the spiral arms and the bridge of stars between the two galaxies.

Nature's version of light pollution, a bright Moon, will also make deep sky objects hard to distinguish. Even at a dark site, a full Moon limits the objects you can see. To make matters even more frustrating, the worst time to view the Moon is when it is full. The greatest detail to be seen on the Moon is along the terminator, the line between the light and shadow. During a full Moon, most detail is washed out by the direct sunlight striking the surface. If there are no planets in the sky, you might be limited to viewing double stars and the brightest deep sky objects.

Finally, remember good observation technique as discussed in Chapter 2. Averted vision, wiggling the scope, and varying the magnification will help you to detect fainter objects. Careful observation of an object will bring out details not easily seen at first glance. Experience at the eyepiece will soon pay off.

Star Charts, Magazines, and Other Printed Resources

When selecting objects for a nightly session, printed references are almost essential. A good star chart shows the sky with stars and deep sky objects annotated. To maximize your time under the stars, you might prepare a list of interesting objects by studying such a chart. While a star chart is a *requirement* when using a non-GoTo scope, when you are out under the night sky it will still prove useful to help you identify unknown objects in the eyepiece. For example, perhaps you are viewing M105 and you notice two other faint objects nearby. A detailed star chart will let you quickly identify them as NGC3389 and NGC3384.

Star charts vary in scale. The star chart typically found in a magazine is a full-sky chart that only shows the planets and the brightest stars and deep sky objects. Good sources for full-sky charts are *Sky and Telescope* magazine, *Astronomy* magazine, and the Internet web site SkyMaps.com. Full-sky charts are fine when you are beginning, but as your experience grows you will seek out fainter objects than are shown on an all-sky chart.

The next step up would be more detailed charts that show sections of the sky and thus more objects. Many introductory astronomy books include such charts, but often they are difficult to use in the field. One exception I have found is the spiral-bound version of Terence Dickinson's *Nightwatch*. It includes fine beginners' charts for the Northern Hemisphere with stars down to about magnitude 6 – naked-eye limit. Hundreds of interesting deep sky objects are plotted on these charts.

For additional targets during your nighttime sessions, try Orion's *Deep Map 600*. This large folding star chart covers the entire sky for Northern Hemisphere observers and a nice portion of the sky visible in the Southern Hemisphere. As the name implies, it charts the position of 600 deep sky objects. Or turn to Wil Tirion's *Bright Star Atlas 2000.0* for another excellent resource for planning your evening. Ten charts map the sky, with stars down to magnitude 6.5 and more than one thousand deep sky objects.

Another must is a good lunar map. You will enjoy the Moon much more with a guide to the craters, mountains, and other surface features. Orion and *Sky and Telescope* both offer fine Moon maps for beginners.

Computer Software

The modern computer has dramatically changed the world of the professional astronomer. Intense calculations that took many hours to perform by hand are now completed in fractions of a second by computer. Computer simulations are used routinely to bolster or discount theories of stellar and cosmic evolution. And naturally, observatory telescopes are generally controlled by computers.

The world of the amateur astronomer has also benefited from the widespread availability of cheap, powerful computers. Huge catalogs of objects that were unwieldy at best are now easily sorted and searched. Sky charts that took years to plot accurately are now effortless and instantly created on the humble home computer. Even the palmtop computer has gotten into the act, plotting stars and deep sky objects on its diminutive screen or producing Moon phases and other ephemeris at a moment's notice.

I categorize astronomy software into three types: planetarium, session planning, and specialized.

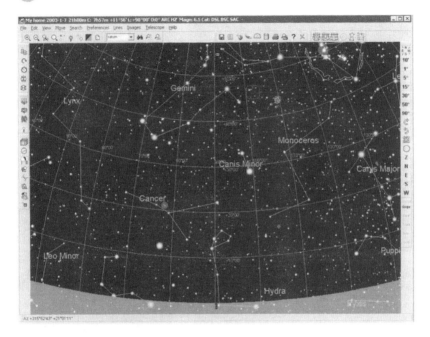

Figure 6.1. Planetarium software – Cartes du Ciel.

Planetarium Software

Planetarium software is the most generic category. Based on several databases of objects, mostly stars and deep sky objects, planetarium software is designed primarily to produce star charts on the computer screen. The charts are plotted based on a location (longitude and latitude or city) and just about any date and time you would like. Objects in the extensive databases are plotted precisely on the charts under the control of filters that determine what appears and what is hidden. Planetarium software is potentially more useful in the field than printed star charts, if you have a suitable portable computer. The view on the computer screen can be made to match the view in any given eyepiece, binoculars, or the naked eye. Printed star charts are limited by their static nature.

There are many good planetarium software packages available, but the three I find most useful for the computer are TheSky, Starry Night Pro, and Cartes du Ciel. All are fine tools for the amateur astronomer and also all will interface with a NexStar telescope to provide enhanced control of both the scope and the software. For palmtop computers TheSky Pocket Edition and Planetarium for Palm are great programs for maximum portability. Both provide support for NexStar control, although at the time of this writing TheSky PE does not support all models. For additional programs and the cost and web sites for these, refer to Appendix C.

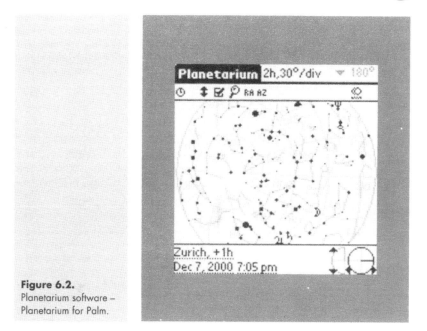

Figure 6.2.
Planetarium software –
Planetarium for Palm.

Session Planning Software

Session planning software emphasizes the ability to search extensive catalogs of objects and produce lists of potential objects for viewing. Most applications also provide sky charting capabilities, but not generally with the feature set of planetarium software. This category of software helps you get organized before you step out for the night. My favorites in this category are NexStar Observer List (my own program), Deepsky 2002, and AstroPlanner. AstroPlanner is especially notable as it is the only planning software currently available for both Macintosh and Windows computers. All three of these interface with NexStar telescopes, allowing you to work down your list, directing your scope to point at one object after another. You will also find that programs of this type have the strongest capabilities in recording and organizing your typewritten observation comments. Refer to Appendix C for web site and cost information.

Specialized Software

Specialized astronomy software is the catchall category. One of my favorites is the Virtual Moon Atlas by Patrick Chevalley and Christian Legrand. Available free for download from http://astrosurf.com/avl/UK_index.html, Virtual Moon Atlas allows you to easily identify hundreds of surface features on the Moon as it provides a very realistic view including the changing phases as the terminator creeps across the lunar surface. You are provided with complete control of the image as you can pan and zoom to zero in on any area of the visible surface of the Moon. This program is not a simple image of the Moon, but rather a fully realistic simulation of the Moon as it actually appears during its ever-changing phases.

Figure 6.3. Session planning software – AstroPlanner.

Another excellent free offering is the AstroByte Logging System by Ron Reuter. AstroByte Logging System allows you to easily record your observations of any of the thousands of objects in its extensive database. You can also display or print a wide variety of reports. Download at http://mainbyte.com/astrobyte.

One last example is Mars Previewer II – a free program by Leandro Ríos. Mars Previewer II displays a detailed view of the surface of Mars for any given date and time. Using this software, you can easily identify the features viewed in the eyepiece. Mars Previewer doesn't seem to have a permanent home, so you might need to search a little for it, but at the time of this writing it is available for download at http://www.astronomysight.com/as/start/books.html#Programs.

For additional software, I keep my eye on advertisements and announcements in monthly magazines like *Sky and Telescope* and *Astronomy*.

The Solar System

Viewing our closest celestial neighbors is the most common activity for beginners, but unfortunately it is common for such viewing to be nothing more than a quick glance before moving on to the next object. Armed with a little knowledge, we find there is much more to be seen.

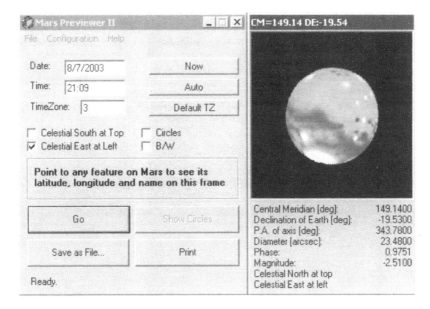

Figure 6.4. Mars Previewer II.

The Sun

One of the most overlooked areas of amateur astronomy is solar observation. While all the other stars we view are nothing more than distant points of light, the Sun provides a full disk view and continually changing detail. Until recently safe solar filters were quite expensive, but such is no longer the case. As we will discuss in Chapter 8, solar filters of excellent quality are now available for a very modest price. A safe filter is a **must**; direct viewing of the Sun with a faulty or inappropriate filter can instantly result in **permanent** eye damage.

After arming your telescope with a good filter, you will find the surface of the Sun is littered with ever-changing sunspots. During the period of maximum solar activity – about every eleven years – the Sun will often be covered with sunspots. Try sketching them every other day for a two-week period to record their evolution and movement. The Sun rotates once about every 30 days, so longer-lived sunspots will be visible for some time. A simple sketch sheet such as the one shown in Figure 2.4 (available from my NexStar Resource Site) makes this enjoyable and easy. Use the circle on the sheet to represent the entire surface of the Sun.

The Moon

Most beginners turn their scope first on the Moon and wonder at this alien landscape. Then, after building the observing skill necessary to enjoy the planets and deep sky objects, the Moon seems to be the last thing on their list. No other object in the sky offers nearly as much detail – and the view changes from one hour to the next. Regardless of the instrument, the Moon is always a stunning sight with even modest magnification. The Moon is an especially rewarding target on evenings when the sky is steady but transparency suffers.

In addition to a telescope, two other tools are a necessity for extensive lunar observing: a lunar atlas or map and a lunar filter. The Moon is incredibly bright, even in a small telescope; the filter reduces the intensity, allowing for much more detail in the eyepiece. Refer to Chapter 8 for advice on selecting a filter. The atlas or map adds much to the visual enjoyment as you put names to the craters and other features, slowly learning the lunar geography. Besides printed maps, consider the Virtual Moon Atlas introduced earlier in this chapter.

Surface features are much easier to detect when they lie along the terminator between the light and dark sides of the Moon. A patient observer can actually watch the terminator slide slowly along the surface as the Sun rises or sets across the lunar landscape. Night after night the terminator illuminates an entirely different area of the Moon.

When viewing the Moon, watch for smaller craters inside of larger craters. Also, many larger craters have small peaks in them. Some craters have lines of debris streaking away, evidence of the impacts that created them. Watch for mountain ranges and other linear features. Many of the longer features have terraced faces visible in moderately sized telescopes. If you have an artistic talent, try sketching the features that you see. Otherwise, keep track of your discoveries, entering them by name in your logbook.

The Planets

The two inner planets – Mercury and Venus – offer little excitement for the Earth-bound observer. Simply identifying Mercury is considered a success for this diminutive planet. At best, Mercury will never be more than about 28° from the Sun. If you decide to use your GoTo telescope to point out Mercury, be absolutely certain the Sun is not in the field of view as well! Refer to finder charts in magazines or your favorite planetarium software to locate the location of this swift traveler. Venus offers a little more variety. Check in on the brightest of planets weekly for two or three months. Sketch the phase and relative size as the view changes slightly from week to week.

Mars ranks number three on my list of planetary favorites. About every other year the red planet approaches close to Earth and every 15 years the pass is dramatically closer. During these close passes Mars presents detail visible in all but the smallest telescopes. 2003 is the best apparition we can expect for many years, with slightly less favorable conditions in 2005. Sketching the view in the eyepiece is one way to record the occasion. You might also try capturing the contrasting surface detail with a digital camera trained on a high-magnification eyepiece.

Jupiter is my favorite planet and one of my overall favorite objects in the sky. In small scopes I enjoy challenging myself to detect a third and sometimes a fourth band on the planet's surface. Watching the orderly march of the four Galilean moons in binoculars is a delight. In a small telescope or binoculars you can watch as one of the moons disappears, hiding in front of or behind giant Jupiter. Viewing such motion, as well as the phases presented by Venus, convinced Galileo that the Sun was the center of our local celestial neighborhood, rather than the Earth as popular wisdom then held. Try sketching the location of the moons every two hours, night after night, and see if you can determine their orbital patterns.

It is in telescopes of about 5-inch aperture or larger that the king of planets really comes to life. In such a scope, watch for detail in the cloud bands – including the Great Red Spot. Jupiter rotates once in a little less than 10 hours, presenting an ever-changing view. A larger scope will also let you enjoy tracking one of the moons as it transits in front of

Jupiter. Or, observe as the inky black shadow of a moon passes across the surface of the planet. An artistic hand is required to due justice to the intricate patterns you will see, or you might try a digital camera. Refer to *Sky and Telescope* magazine or their web site (http://www.SkyandTelescope.com) for the dates and times of events on Jupiter or see if your planetarium software provides such details.

Saturn places a close second to Jupiter in my book. Visually it is certainly the most beautiful sight in the sky, but the greater detail to be had when viewing Jupiter holds my interest longer. The challenge and enjoyment of Saturn lies in the subtle details to be found in the rings, the planet disk, and the faint moons. In a small telescope, look for the Cassini Division – the dark line between the two brightest rings – on the nights of best seeing. Look for the shadow of the planet disk on the rings behind the planet. Only Titan, Saturn's brightest moon, is easily noticeable in a small scope.

In an 8-inch or larger scope, look for a clearly defined Cassini Division as well as the fainter C ring (Crepe ring) inside the other two. Exceptional nights of clear air will also allow faint bands to be viewed on the planet disk. Five or more moons are within reach of large telescopes. The bright golden hues of Saturn are a good target for a digital camera.

The three outer planets, Uranus, Neptune, and Pluto, offer the same kind of challenge as Mercury. Locating and identifying them in the eyepiece is about all we can expect to accomplish. Uranus and Neptune are fairly easy objects in a moderately sized scope. Both present a disk, Neptune substantially fainter than Uranus, in a high-power view. See if you can make out any color in their disks. Pluto is much more challenging. Pluto is an extremely dim magnitude 14 and so tiny that it doesn't present a disk, only a faint point of light. On extremely clear nights, an 8-inch scope can just make it out. Even with GoTo to point directly at this tiny wanderer, you will have quite a hard time determining which faint point of light is the object of your desire! Try sketching the view in the eyepiece every night for a week; the moving point of light is Pluto.

Deep Sky Objects

While the solar system provides us with a short list of some of astronomy's most exciting celestial views, deep sky objects – DSOs – are so numerous we can view several different objects every night of the year. With hundreds of thousands of galaxies, nebulae, and star clusters visible in amateur telescopes, the variety is remarkable. The challenge is settling on a list of targets that will prove interesting and enjoyable for the evening.

Learning the constellations will help, as that is the most practical way to explore the sky – a section at a time. You should also keep in mind our discussion of magnitude and surface brightness from earlier in this chapter. Even though a GoTo scope can point directly at any object you request, that won't make it visible in the eyepiece. Magazines and club newsletters often highlight lesser-known DSOs of interest, but most often we turn to several established lists.

The Messier Catalog

By far the most well-known list of DSOs is the Messier catalog. A collection of 109 (110 from most sources) of the best DSOs for the Northern Hemisphere, the Messier list presents a wonderful variety of objects to get you started in deep sky astronomy. A common endeavor in amateur astronomy is to log the viewing of all objects in the Messier list. All are visible in even the humble 60mm refractor, but only under clear, dark skies. Under less favorable conditions substantially more aperture is called for.

Table 6.2 shows the Messier catalog, organized by season for best viewing. I have rated each object with a difficulty rating:

- **Easy** Easily visible with any aperture even in most city skies.
- **Medium** Easily visible with small apertures (60–80mm) under dark skies, just visible in moderate apertures (4–5 inches) in suburban skies, visible in larger apertures (8-inch or more) in most city skies.
- **Hard** Just visible with small apertures under dark skies, easily visible with moderate aperture under dark skies, just visible in large apertures in suburban skies.

Note that this rating system expects great conditions and indicates that you can see the object, not necessarily any detail. Great conditions include a night of good transparency with no Moon and the object is high in altitude – not possible for all objects from some northern latitudes. In particular, northern observers will have difficulty with the objects in Sagittarius, Scorpius, and other low-lying constellations.

Double Stars

Double stars are quite beautiful and a good target when sky conditions are unfavorable for most other deep sky objects. Doubles hold up well when a bright Moon washes out most other objects, and many don't require any better-than-average-seeing. The current NexStar models identify 55 double stars in the hand control. Table 6.3 is a list of these doubles organized by season. Refer to Chapter 3 for the resolving limit of each NexStar scope to determine the likelihood of splitting any given double star.

For more double stars, refer to the Downloads section of my NexStar Resource Site (http://www.NexStarSite.com). The star lists available include hundreds of additional double stars accessible via the **Star** button on the hand control.

The Caldwell List

The Caldwell list is another great resource for deep sky objects. None of the objects are included in the Messier list, so it is common for amateur astronomers to work their way through both lists concurrently. Comprised of 109 objects published in a *Sky and Telescope* article by Patrick (Caldwell-)Moore, the list includes some of the most spectacular DSOs not catalogued by Charles Messier. In particular you will find the open clusters and planetary nebulae to be among the best in the sky. Additionally, the Caldwell list includes objects visible only in the Southern Hemisphere, an area of the sky not experienced by Messier. All 109 objects are accessed via the **Cald** button on the hand control.

The official list can be found on *Sky and Telescope*'s web site (http://www.skyandtelescope.com) or you can download a list suitable for sorting in the Downloads section of my NexStar Resource Site (http://www.NexStarSite.com).

The Herschel 400

Perhaps the best list of DSOs after completing the Messier and Caldwell catalogs, the Herschel 400 list was compiled by the Ancient City Astronomy Club to provide continuing

Table 6.2. The Messier catalog

	Object number	Name	Type	Difficulty	Constellation
Spring	46	NGC 2437	Open Cluster	Medium	Puppis
	47	NGC 2422	Open Cluster	Easy	Puppis
	93	NGC 2447	Open Cluster	Medium	Puppis
	50	NGC 2323	Open Cluster	Easy	Monoceros
	44	NGC 2632 Beehive Cluster (Praesepe)	Open Cluster	Easy	Cancer
	67	NGC 2682	Open Cluster	Medium	Cancer
	48	NGC 2548	Open Cluster	Medium	Hydra
	81	NGC 3031 Bode's Galaxy	Spiral Galaxy	Medium	Ursa Major
	82	NGC 3034 Cigar Galaxy	Irregular Galaxy	Medium	Ursa Major
	97	NGC 3587 Owl Nebula	Planetary Nebula	Medium	Ursa Major
	108	NGC 3556	Spiral Galaxy	Hard	Ursa Major
	109	NGC 3992	Spiral Galaxy	Hard	Ursa Major
	40	Winecke 4	Double Star	Easy	Ursa Major
	95	NGC 3351	Spiral Galaxy	Hard	Leo
	96	NGC 3368	Spiral Galaxy	Hard	Leo
	105	NGC 3379	Elliptical Galaxy	Hard	Leo
	65	NGC 3623	Spiral Galaxy	Medium	Leo
	66	NGC 3627	Spiral Galaxy	Medium	Leo
	94	NGC 4736	Spiral Galaxy	Medium	Canes Venatici
	106	NGC 4258	Spiral Galaxy	Hard	Canes Venatici
	64	NGC 4826 Blackeye Galaxy	Spiral Galaxy	Medium	Coma Berenices
	85	NGC 4382	Lenticular (S0) Galaxy	Hard	Coma Berenices
	88	NGC 4501	Spiral Galaxy	Hard	Coma Berenices
	91	NGC 4548	Spiral Galaxy	Hard	Coma Berenices
	98	NGC 4192	Spiral Galaxy	Hard	Coma Berenices
	99	NGC 4254	Spiral Galaxy	Hard	Coma Berenices
	100	NGC 4321	Spiral Galaxy	Hard	Coma Berenices
	68	NGC 4590	Globular Cluster	Hard	Hydra

Table 6.2. The Messier catalog (continued)

Object number	Name	Type	Difficulty	Constellation
49	NGC 4472	Elliptical Galaxy	Hard	Virgo
58	NGC 4579	Spiral Galaxy	Hard	Virgo
59	NGC 4621	Elliptical Galaxy	Hard	Virgo
60	NGC 4649	Elliptical Galaxy	Medium	Virgo
61	NGC 4303	Spiral Galaxy	Hard	Virgo
84	NGC 4374	Lenticular (S0) Galaxy	Hard	Virgo
86	NGC 4406	Lenticular (S0) Galaxy	Hard	Virgo
87	NGC 4486 Virgo A	Elliptical Galaxy	Hard	Virgo
89	NGC 4552	Elliptical Galaxy	Hard	Virgo
90	NGC 4569	Spiral Galaxy	Hard	Virgo
104	NGC 4594 Sombrero Galaxy	Spiral Galaxy	Medium	Virgo
Summer				
3	NGC 5272	Globular Cluster	Easy	Canes Venatici
51	NGC 5194 Whirlpool Galaxy	Spiral Galaxy	Medium	Canes Venatici
63	NGC 5055 Sunflower Galaxy	Spiral Galaxy	Hard	Canes Venatici
53	NGC 5024	Globular Cluster	Medium	Coma Berenices
83	NGC 5236 Southern Pinwheel Galaxy	Spiral Galaxy	Medium	Hydra
101	NGC 5457 Pinwheel Galaxy	Spiral Galaxy	Hard	Ursa Major
102	NGC 5866 Spindle Galaxy	Lenticular (S0) Galaxy	Medium	Draco
5	NGC 5904	Globular Cluster	Easy	Serpens Caput
57	NGC 6720 Ring Nebula	Planetary Nebula	Easy	Lyra
13	NGC 6205 Hercules Globular Cluster	Globular Cluster	Easy	Hercules
92	NGC 6341	Globular Cluster	Easy	Hercules
9	NGC 6333	Globular Cluster	Medium	Ophiucus
10	NGC 6254	Globular Cluster	Medium	Ophiucus
12	NGC 6218	Globular Cluster	Medium	Ophiucus
107	NGC 6171	Globular Cluster	Medium	Ophiucus
14	NGC 6402	Globular Cluster	Medium	Ophiucus
19	NGC 6273	Globular Cluster	Easy	Ophiucus

Table 6.2. The Messier catalog (continued)

Object number	Name	Type	Difficulty	Constellation
62	NGC 6266	Globular Cluster	Medium	Ophiucus
4	NGC 6121	Globular Cluster	Medium	Scorpius
80	NGC 6093	Globular Cluster	Medium	Scorpius
6	NGC 6405 Butterfly Cluster	Open Cluster	Easy	Scorpius
7	NGC 6475 Ptolemy's Cluster	Open Cluster	Easy	Scorpius
23	NGC 6494	Open Cluster	Easy	Sagittarius
8	NGC 6523 Lagoon Nebula	Diffuse Nebula	Medium	Sagittarius
17	NGC 6618 Swan Nebula	Diffuse Nebula	Medium	Sagittarius
18	NGC 6613	Open Cluster	Easy	Sagittarius
20	NGC 6514 Trifid Nebula	Diffuse Nebula	Medium	Sagittarius
21	NGC 6531	Open Cluster	Easy	Sagittarius
22	NGC 6656	Globular Cluster	Easy	Sagittarius
24	NGC 6603 Milky Way Patch	Star Cloud	Easy	Sagittarius
25	IC 4725	Open Cluster	Easy	Sagittarius
28	NGC 6626	Globular Cluster	Medium	Sagittarius
54	NGC 6715	Globular Cluster	Medium	Sagittarius
69	NGC 6637	Globular Cluster	Medium	Sagittarius
70	NGC 6681	Globular Cluster	Medium	Sagittarius
11	NGC 6705 Wild Duck Cluster	Open Cluster	Easy	Scutum
26	NGC 6694	Open Cluster	Easy	Scutum
16	NGC 6611 Eagle Nebula Cluster	Open Cluster	Easy	Serpens Claudia
Fall				
55	NGC 6809	Globular Cluster	Medium	Sagittarius
75	NGC 6864	Globular Cluster	Medium	Sagittarius
71	NGC 6838	Globular Cluster	Hard	Sagitta
27	NGC 6853 Dumbbell Nebula	Planetary Nebula	Medium	Vulpecula
56	NGC 6779	Globular Cluster	Medium	Lyra
2	NGC 7089	Globular Cluster	Easy	Aquarius
72	NGC 6981	Globular Cluster	Hard	Aquarius

Table 6.2. The Messier catalog (continued)

	Object number	Name	Type	Difficulty	Constellation
	73	NGC 6994 Group of 4 stars	Group/Asterism	Easy	Aquarius
	29	NGC 6913	Open Cluster	Easy	Cygnus
	39	NGC 7092	Open Cluster	Easy	Cygnus
	30	NGC 7099	Globular Cluster	Medium	Capricornus
	15	NGC 7078	Globular Cluster	Easy	Pegasus
	52	NGC 7654	Open Cluster	Medium	Cassiopeia
	31	NGC 224 Andromeda Galaxy	Spiral Galaxy	Easy	Andromeda
	32	NGC 221 Satellite of M31	Elliptical Galaxy	Medium	Andromeda
	110	NGC 205 Satellite of M31	Elliptical Galaxy	Hard	Andromeda
Winter	103	NGC 581	Open Cluster	Easy	Cassiopeia
	76	NGC 650 Little Dumbbell / Cork Nebula	Planetary Nebula	Medium	Perseus
	34	NGC 1039	Open Cluster	Easy	Perseus
	74	NGC 628	Spiral Galaxy	Hard	Pisces
	77	NGC 1068 Cetus A	Spiral Galaxy	Hard	Cetus
	36	NGC 1960	Open Cluster	Easy	Auriga
	41	NGC 2287	Open Cluster	Easy	Canis Major
	35	NGC 2168	Open Cluster	Easy	Gemini
	33	NGC 598 Triangulum/Pinwheel Galaxy	Spiral Galaxy	Hard	Triangulum
	37	NGC 2099	Open Cluster	Easy	Auriga
	38	NGC 1912	Open Cluster	Easy	Auriga
	79	NGC 1904	Globular Cluster	Medium	Lepus
	42	NGC 1976 Great Orion Nebula	Diffuse Nebula	Easy	Orion
	43	NGC 1982 de Mairan's Nebula	Diffuse Nebula	Medium	Orion
	78	NGC 2068	Diffuse Nebula	Medium	Orion
	45	Pleiades, Subaru, Seven Sisters	Open Cluster	Easy	Taurus
	1	NGC 1952 Crab Nebula	Supernova Remnant	Medium	Taurus

Table 6.3. Double stars identified in the current NexStar models

	Name	Magnitude of primary	Separation	Constellation
Spring	145 Cma	4.8	27″	Canis Major
	Castor	1.6	1.8″	Gemini
	19 Lyn	5.6	14.8″	Lynx
	Eta Puppis	5.9	10″	Puppis
	Kappa Puppis	4.6	10″	Puppis
	Tegman	6.0	9″	Cancer
	Theta 2 Cnc	6.3	5″	Cancer
	Iota Cancer	4.2	31″	Cancer
	Algieba	2.6	4.4″	Leo
	54 Leo	4.5	6.8″	Leo
	Cor Caroli	2.9	19″	Canes Venatici
	24 Com	5.2	20″	Coma Berenices
	35 Com	5.1	29″ (Triple)	Coma Berenices
	Algorab	3.1	24″	Corvus
	Porrima	2.9	3″	Virgo
Summer	Mizar	2.4	14″	Ursa Major
	Kappa Boo	4.6	13″	Bootes
	Epsilon Boo	2.7	3″	Bootes
	Xi Boo	4.6	7″	Bootes
	Delta Ser	5.2	3.9″	Serpens
	Graffias	2.9	14″	Scorpius
	Nu Dra	5.0	62″	Draco
	Rasalgethi	3.5	4.6″	Hercules
	95 Her	5.1	6.5″	Hercules
	Epsilon Lyr 1	6.0	2″	Lyra
	Epsilon Lyr 2	4.5	2.2″	Lyra
	Zeta Lyr	4.3	44″	Lyra
Fall	Albireo	3.2	35″	Cygnus
	17 Cyg	5.0	26″	Cygnus
	Dabih	3.3	205″	Capricornus
	Omicron Cap	6.1	19″	Capricornus
	61 Cyg	5.4	30″	Cygnus
	Epsilon Peg	2.5	83″	Pegasus
	Zeta Aqr	4.6	2″	Aquarius
	Tau 1 Aqr	5.7	23″	Aquarius
	Delta Cep	4.0	20″	Cepheus
	94 Aqr	5.3	13″	Aquarius
	107 Aqr	5.8	7″	Aquarius
	Sigma Cas	4.9	3″	Cassiopeia
	Eta Cas	3.6	12″	Cassiopeia
Winter	Gamma Aries	4.8	8″	Aries
	Lamda Aries	4.8	38″	Aries
	Psi Piscium	5.5	30″	Pisces
	Zeta Piscium	5.6	23″	Pisces
	Almach	5.1	10″	Andromeda
	30 Ari	6.6	39″	Aries

Table 6.3. Double stars identified in the current NexStar models *(continued)*

Name	Magnitude of primary	Separation	Constellation
Gamma Cet	3.6	2.7"	Cetus
Acamar	4.4	8"	Eridanus
Polaris	2.1	18"	Ursa Minor
32 Eri	5.0	7"	Eridanus
Rigel	0.3	9"	Orion
Mintaka	2.5	53"	Orion
Adhara	1.6	7.5"	Canis Major
38 Gem	4.7	7"	Gemini
Beta Mon	4.7	3"	Monoceros

challenges for intermediate-level amateur astronomers. Choosing the best objects from a catalog compiled by William Herschel in the late eighteenth century, they compiled a challenging list of a wide variety of DSOs. The entire list can be viewed and printed from the Astronomical League's web site (http://www.astroleague.org/al/obsclubs/herschel/fwhershs.html) or a list suitable for sorting can be had in the Downloads section of my NexStar Resource Site (http://www.NexStarSite.com).

The Dunlop 100

Observers in the Northern Hemisphere compiled most of the lists mentioned thus far. Many of the objects in these lists are not visible from the Southern Hemisphere. In the early nineteenth century James Dunlop recorded observations of 629 objects from Australia. Recently Glen Cozen compiled the best 100 of these objects into a catalog now commonly known as the Dunlop 100. The list can be found at http://www.seds.org/messier/xtra/similar/dunlop.html – the web site of the Students for the Exploration and Development of Space.

Sources of Other Deep Sky Object Lists

The SEDS web site is also the best resource for additional catalogs of DSOs. Visit the web page http://www.seds.org/messier/xtra/similar/catalogs.htm for many more catalogs, most compiled by amateur astronomers for use with modest equipment.

Observation Programs

If deciding what to seek out on your own sounds like too much work, there are alternatives. Start by checking with your local astronomy club to see what observation programs they offer. Most clubs have planetary, lunar, double star, variable star, and deep sky observation programs to keep their members busy. Each is generally a list of objects or objectives for the observer to complete. Such programs can keep you busy for years and provide concrete goals to keep your interest peaked.

Figure 6.5. NexStar
50 Club logo.

If you do not have a convenient local club, check into the Astronomical League observing programs at their web site – http://www.astroleague.org. Here you will find observing programs ranging from naked-eye observation of the Moon to advanced deep sky projects. All are well thought out and provide good suggestions to get you started. Upon completion, members of the Astronomical League receive a certificate. Be forewarned that for many of these programs certificates are not awarded if a computerized telescope is used.

One Internet-based club that will award certificates when a computerized telescope is used is the NexStar 50 Club (http://www.NexStarSite.com/nexstar50club.htm). In fact, it is a requirement that a NexStar telescope is used to observe all objects! The club was an idea generated by the Yahoo NexStar Group (http://groups.yahoo.com/group/nexstar) and is administered by Hank Williams and myself. Membership is granted to anyone using a NexStar telescope to view at least 40 of the 50 objects in the NexStar 50 List. The 50 objects are various solar system and deep sky objects voted as favorites by members of the NexStar Group. Visit the site and see which objects you have already logged; maybe you will be the next person I send a certificate to!

Chapter 7
Using the Software Included with Little NexStars – TheSky and NexStar Observer List

As discussed in the last chapter, a good tool for deciding what you would like to observe is planetarium software. The little NexStars (60/80/114/4) include a CD with TheSky Level I from Software Bisque. Other models of NexStars have occasionally included TheSky as well. While this program can be very useful in planning an observing session, there are so many features that it is somewhat confusing to use. A complete user's guide is copied to your computer during installation, but most do not seem to spend much time reading the more than ninety pages provided. This chapter will present a clear overview of the most useful features with a much shorter reading requirement.

The little NexStars also ship with a second program, NexStar Observer List (NSOL). Owners of other NexStar telescopes can download the program at no charge from http://www.NexStarSite.com. It just so happens that the author of that program is the author of this book! NSOL allows your personal computer (PC) to control your NexStar telescope and helps you to keep organized during the precious time you spend under clear skies. In this chapter we will explore just what NSOL is and how to make practical use of it.

TheSky

TheSky is considered by many to be the premier planetarium software available today. It provides very accurate screen and printed sky charts, includes hundreds of thousands (even millions in some versions) of objects in an extensive database, plus thousands of beautiful photographic images, and allows very comprehensive control of what will be displayed.

Software Bisque offers TheSky in several different configurations or "levels". This allows you to purchase the capabilities you need, without paying for features you will not use. Level I is not generally offered for sale, but rather is packaged for distribution with some telescopes such as the little NexStars. Level I includes stars down to magnitude 11 and the entire NGC, IC, and Messier catalogs. Level I can be upgraded to Level II for $49, much less than the normal $129 price tag. It is important to note that Level I does **not** provide telescope control. Levels II and higher allow control of NexStar telescopes. Visit Software Bisque's web site – http://www.bisque.com – for information on Levels II and higher. Submit a question via their support page to inquire about upgrading to Level II.

The rest of this section is devoted to TheSky Level I, although most topics apply equally to all levels. Menus, names on buttons and other objects found on the screen are shown in **bold** text. Buttons on the toolbars are shown next to discussions of their function. Keys on the computer keyboard are shown in ***bold italics***.

Installation

Installation is very simple following the on-screen instructions. If the installation program doesn't start after inserting the CD, open **My Computer**, double-click the CD drive, and then double-click the **Setup** icon. After acknowledging the location for installation and the license agreement, you will be prompted to select one of three **Setup Types**: Compact, Custom, or Typical. If you have more than 300MB of hard drive space to devote to TheSky, I recommend installing the entire contents of the CD to the hard drive – this is the Custom option. If you do not install the entire CD, it is necessary to insert the CD in the computer each time **before** starting TheSky. If you select Custom, you will be prompted to select which components to install. All are selected except the optional Astronomy Tutorial, which is not required to make full use of TheSky.

Initial Configuration of TheSky

TheSky offers many settings to control exactly what you see on the screen. Many of these options we change repeatedly but there are some settings we configure occasionally and seldom change. Here are the settings I recommend you configure before you start using TheSky:

- **Your Location** Found on the **Data** menu, **Site Information** is used to tell TheSky where you are on the Earth and the date and time it should use to construct the display. On the **Location** sheet of the dialog box, **Description** provides a list of cities in the United States for easy selection of your location. Use the **Open** button to select alternate files containing locations in other countries, such as "Cities outside USA". If you do not live near any of the cities listed, select one in your time zone and enter your longitude, latitude, and elevation directly. If you have a GPS receiver or a NexStar GPS telescope, you can determine your location to a high degree of precision. Otherwise, you can find your location at the Heavens Above web site – http://www.heavens-above.com/countries.asp. A local airport can provide your elevation, although you do not need a great deal of precision to produce good results. You can also name your location. To do so, complete the longitude, latitude, and other information, then type the name in the Description box. Then click the **Add** button. In this way, you can save several locations where you commonly observe the night sky and switch between them using the **Description** list.

- **Date and Time** Also found in **Data** menu, **Site Information**, the **Date and Time** sheet allows you to set the date back to 4713 BC or forward to 10 000 AD. When using TheSky out under the night sky, you will generally want to check the box to **Use computer's clock**. Also note you must correctly set the daylight savings time option for your location.

- **Display Date and Time in the Status Bar** The Time Skip operations we will explore later in this chapter cause TheSky to "fast-forward" or "rewind" through time. As this occurs you generally would like to know the current date and time for the view being displayed. TheSky can display the current date and time in the **Status Bar** at the bottom of the screen. Select **Status Bar** from the **View** menu and select the date and time options.

- **Horizon Color** The default color for the area below the horizon is green on the screen and red on printed charts. Green proves to be too bright when using TheSky at night as it diminishes your night vision. Red wastes enormous amounts of ink or toner on printed charts. To change these colors, select **Preferences** from the **View** menu. Under **Object Description**, scroll down to **Horizon Lines** then click the **Fill** button. The color shown in the **Virtual Sky Sample** box is the color used on the screen, while **Sky Chart Sample** is the color used on printed charts. Click each **Edit** button to select more appropriate colors. For the Virtual Sky, I select a very dark maroon. For the Sky Chart, I use white.

- **Displaying Pop-Up Labels for Objects on the Screen** We will discover later that for objects with common names, such as the brighter stars or the Messier objects, we can display labels next to the objects. Other objects cannot be identified this way. We can configure TheSky to display a pop-up label for objects when you point at the object with the mouse cursor. Select **Toolbars** from the **View** menu and select the **Show ObjectTips** option. If you now point at an object such as a star, a pop-up box will provide its designation.

Screen Modes

TheSky calls the display of the sky the "Virtual Sky". There are several basic options for controlling how the Virtual Sky is displayed on the screen. All of these options are found on the **View** menu:

Spectral Colors Displays stars in colors to approximate their actual visual color. While this mode is useful on the screen, I don't recommend it when printing charts due to the grayscale colors used with black-and-white printers.

 Daytime Sky Mode Displays the sky as blue when the Sun is above the horizon. As time progresses and the Sun sets, more and more stars become visible, much as is experienced in the real sky.

 Night Vision Mode Changes the display to various dark-red tones to preserve your night vision. Night Vision Mode does not affect the Virtual Sky and the default color of the horizon (green) is much too bright. Refer to the previous section for details on selecting a more friendly color for the horizon.

Chart Mode A print preview mode displaying the sky with a white background and mostly black objects.

 Full Screen Causes the Virtual Sky to expand to fill the entire screen. Press the *Spacebar* on the keyboard to toggle between normal and full screen modes. The **Show in Full Screen Mode** option on **View, Toolbars** controls whether the various toolbars will be displayed in full screen mode.

Mirror Image Flips the Virtual Sky left to right. This can be useful when comparing the view in the eyepiece of a refractor, SCT, or Maksutov to the display on the computer screen.

Moving Around the Virtual Sky

TheSky offers several methods for moving the view in the Virtual Sky. The four arrow or **Move** buttons on the toolbar offer one method, or you can use the scroll bars on the edge of the screen for the same basic effect. The arrow keys on the keyboard can also be used to move the display left, right, up, or down. To undo as many as the last 15 move changes, press *Ctrl+Z* (⌘ *+Z* on the Macintosh). To scroll the Virtual Sky to center a specific point, right-click (Control-click on the Macintosh) that point and select **Center**.

TheSky has three orientation modes that determine how the Virtual Sky scrolls when you direct it to move. All three are found on the **Orientation** menu:

 Zenith Up The zenith – the point in the real sky directly above your head – is located at the top of the screen. When the zenith is not visible in the Virtual Sky, imagine that it is locked in position just above your computer's monitor. Use this mode most of the time unless you are zooming in on an area of the sky and wish to determine orientation based on the celestial pole (common for studying double stars and other objects located close together). Left and right movements are as if you are facing the horizon and turn left and right. Up and down movements are as if you are moving your gaze straight up and down.

 Pole Up The celestial north pole – for those in the Northern Hemisphere – is located at the top of the screen. Naturally the celestial south pole is at the top for those in the Southern Hemisphere. Again, if the pole is not visible in the Virtual Sky, consider it locked in position about the screen. In this mode the horizon disappears from the screen as you are generally studying a small section of sky without regard to the local horizon. Moving to the left increases right ascension and moves the Virtual Sky the same direction as the rotation of the Earth. Moving to the right decreases right ascension. Moving up and down changes declination. While all of this seems complicated, remember that you normally use **Pole Up** orientation only when examining a small area of the sky. In this case, the movements will seem natural.

 Free Rotation Allows the display to be rotated around the center of the Virtual Sky. Generally used when you have zoomed in to a highly magnified view and are attempting to match the display to the view in the eyepiece of a telescope. To rotate the display, press *Alt+R* (⌘ *+R* on the Macintosh) to display a line with a small circle indicating the direction the North Celestial Pole. Click and drag the circle to rotate the display. You can also use *Alt++* (*Alt* and the "=" key) to rotate the display clockwise and *Alt+-* (Alt and the "-" key) to rotate the display counterclockwise.

In addition to using the **Move** buttons and various other methods discussed, TheSky also offers buttons to cause the Virtual Sky to turn directly to any of the cardinal points – north, south, east, and west – or directly overhead – the zenith. A single click of one of the following **Look** buttons is all that is needed:

You will also note that the screen zooms out to a field of view of 100° – approximately the view seen with the naked eye.

Zooming In and Out

  The most obvious way to zoom in or out on the Virtual Sky is to use the magnifying glass buttons on the toolbar. You can also zoom in or out using *Page Up/Page Down* on the keyboard. Yet another method is to right-click (Control-click on the Macintosh) the mouse anywhere over the Virtual Sky, point at **Zoom To**, and select one of the seven preset fields of view. Note the keyboard shortcuts shown on the **Zoom To** menu such as *Ctrl+T* to zoom to 1°, the approximate field of view in a telescope. The keyboard shortcuts can be used without accessing the menu. The highest magnification allowed is a field of 1 arc second.

One of the most flexible methods (no pun intended) for zooming is the rubber band box, or as Software Bisque likes to call it, the zoom box. Decide the section of the sky you want displayed. Point the mouse in the upper left corner of this section then click and drag to the lower right. As you do, a red "rubber band box" will be drawn around the area. Release the mouse button, point anywhere inside the box, and click. The Virtual Sky will zoom to display just that section of sky – with necessary adjustments to fit the aspect ratio of the Virtual Sky window. Besides giving complete control of the magnification of the display, the zoom box allows you to center the area while zooming. Other methods will usually leave you scrolling your target back to center after zooming to the desired magnification.

In the bottom left of the window, in the status bar, TheSky displays the field of view (FOV) currently displayed. This FOV is actually a measurement from the left to the right edge of the Virtual Sky. To more closely match the view in the eyepiece you should resize TheSky's window to display the Virtual Sky as a square; thus the FOV would indicate both the width and the height of the current display.

As you zoom out, only the brightest objects are displayed in order to minimize screen clutter. As you zoom in, fainter objects are added to the display and some objects such as galaxies, star clusters and nebulae are displayed as outlined shapes indicating the actual size and shape of the objects.

Controlling What Objects Are Displayed – Filters

As noted earlier, the database of objects available in TheSky includes hundreds of thousands and in some versions even millions of objects. Many of these objects may not be visible in your telescope or they may be of little interest to you. The basic tool to control what objects will be included in the display is known as the **Filters** dialog. You can access **Filters** by right-clicking (Control-click on the Macintosh) the mouse anywhere over the Virtual Sky and selecting **Filters**. **Filters** is also found on the **View** menu.

Objects are separated into several categories, giving a high level of control over what will be displayed. Clear the checkboxes of object types you are not interested in displaying on the screen and mark the checkboxes of the objects you want displayed. You can select categories of objects using the six buttons in the lower right and then mark or clear all with one operation.

Filters also provides controls to set the limiting magnitude of each type of object. For example, you can select the star types and set the **Faintest** magnitude to 10. For Galaxies and Nebula, it might be more appropriate to set the limiting magnitude as 8. In this way, you can limit the objects displayed to those most appropriate to your equipment and viewing conditions.

The bottom part of the list is not actually comprised of celestial objects, but rather is a list of features in TheSky. For example, if you are annoyed (as I am) by the large images that pop up when you click on an object, remove the check mark next to **Image**. The images are still available in the information dialog that will now be displayed for each object.

Note that when you are displaying a large area of the sky, not all objects meeting filter criteria will be displayed. Consider that if you set the limiting magnitude to 15 and select all object types, the Virtual Sky would be a solid block of various colors since literally tens of thousands of objects would be found in a 50° patch of sky. As you zoom in to higher magnifications, fainter objects will be displayed down to the limits of your filter settings or the limits of the database.

In addition to the Filters dialog, you can quickly display and hide objects using the following four toolbar buttons:

 Stars

 Star Clusters (Open and Globular)

 Galaxies

 Nebulae

Various lines can be controlled with the **Filters** dialog or via **Reference Lines** on the **View** menu. You will likely find it useful to display constellation boundaries and figures while you are learning your way around the night sky. In addition to these dialog boxes, the following toolbar buttons provide instant control for the most commonly used lines:

 Equatorial Grid Lines of right ascension and declination

 Constellation Figures

 Horizon Grid Lines of azimuth and altitude

 Constellation Borders

To make any of these filter settings "permanent", you must save the current Sky Document as described in a later section.

Comets and Minor Planets

By default TheSky displays just a few of the possible comets and minor planets (asteroids) that are commonly visible in our solar system. In addition to selecting Comet and Minor Planet in the Filter dialog box, you must also tell TheSky which of these types of objects you want to display. This is done in the **Comets and Minor Planets** dialog found on the **Data** menu. Simply check the box next to the objects you wish to display on the Virtual Sky.

As new comets and minor planets are discovered, you can add them to TheSky's database. The easiest way to do this is to take advantage of Internet services created for just this purpose. It couldn't be easier. First, be sure you are connected to the Internet. Next open the **Comets and Minor Planets** dialog from the **Data** menu. To download new comets, click the **Observable** button on the **Comets** sheet. After a short delay, you will be presented with a list of new comets available. Simply select the comet or comets you wish to add. To download new minor planets, click the **Distant, Critical,** or **Unusual** buttons on the **Minor Planets** sheet, and then select from the downloaded list.

Use **Find** from the **Edit** menu to center on a comet or minor planet. Select **Comet** or **Minor Planet** from the **Common Names** section of the **Find** dialog and you will be presented with a list of comets and minor planets to choose from.

Labels

TheSky can display labels next to certain types of objects. The **Labels** button on the toolbar quickly toggles them on or off. Use the **View** menu, **Labels, Setup** to determine which labels are displayed. Turn on constellations to help you learn your way around the sky. If you are interested in locating asteroids and other minor planets, select that option. To identify objects without labels you must click on them to display the **Object Information** dialog box or choose **Toolbars** from the **View** menu and select the **Show ObjectTips** option. ObjectTips will display a pop-up box designating any object you point at in the Virtual Sky.

As mentioned earlier, not all objects can have a label displayed next to their symbol. For example, the thousands of galaxies in the NGC and IC catalogs cannot normally be labeled in TheSky Level I. But there is a trick that will configure Level I to display both NGC and IC labels. The **Find** dialog box, described fully below, can display a view of the Virtual Sky that is equivalent to any of the charts in the printed sky atlas *Uranometria 2000.0*. Simply type "URAnnn+" (where nnn is the Uranometria chart number, for example URA001+ is the chart of the north celestial polar region) in the **Find** box and the Virtual Sky will be configured to display labels next to NGC and IC objects. This also displays the Equatorial Grid, which can be controlled as discussed earlier.

Finding Objects and Displaying Information on Objects

To quickly locate any of the thousands of objects in the database, use the **Find** dialog box. Access find with the "Sherlock Holmes" button on the Toolbar (yep, that's a fellow with a pipe and magnifying glass) or simply press the *F* on the keyboard (⌘ +*F* on the Macintosh). The top half of the find dialog allows easy selection of many common objects. The middle section allows selection from the various catalogs comprising TheSky's complete database. The entry box labeled **Find** allows direct keyboard entry, for example, M31 or NGC7009.

After selecting an object or typing a name, click either the **Find** button or the **Center & Frame** button. The **Find** button displays the **Object Information** dialog box describing the object, while the **Center & Frame** button scrolls the Virtual Sky to the object and zooms in, filling the display with the object. Most likely you will want to display the **Object Information** dialog.

The **Object Information** dialog box will show various detailed information about the object. You will find the object's coordinates, magnitude, alternative names, and other information. Buttons along the bottom of the dialog box allow you to jump directly to certain pieces of data and display more or less information. The **Multimedia** sheet will often display an image of the object or spectral class information for stars. Compare Betelgeuse and Bellatrix to appreciate spectral class.

You may find the following buttons useful:

 Center Centers the object in the Virtual Sky. Does not affect the zoom level. If the object is below the horizon, switch to Pole Up orientation or zoom in until it is displayed.

 Observer's Log Opens a log file (by default a file named log.txt in the *user* subfolder), copies the object information into the end of the file, and allows typed entries. Unfortunately the log file is simply a running log with no sort or filter capabilities. Also, you would need to open the file in another program, such as your word processor, to print.

You can also display the **Object Information** dialog box for any object on the screen by simply clicking on it. If you find it difficult to select the correct object, zoom in and try again. This is the default method of identifying an unlabeled object in TheSky.

Determining the Separation and Position Angle Between Two Objects

The **Object Information** dialog can also be used to measure the angular distance (separation) and orientation (position angle) between two objects. Click the first object to display the dialog box and then click the second object. For example, click on Rigel, then Betelgeuse. Scroll down to the bottom of the information window and you will discover that Betelgeuse is about 18° from Rigel at a position angle of 33°. Position angle is measured counterclockwise starting at a line drawn from the first object to the north celestial pole.

Time Skip Animation

The Time Skip feature is great for simulating the passage of time and how it affects objects in the sky. For example, you might like to track the movement of the Moon or a planet across several nights throughout a month. Or perhaps you would like to observe how the constellations march across the sky night after night. Following are the buttons on the toolbar that control Time Skip.

Skip Increment Specifies the amount of time assigned to each step during Time Skip. In addition to the entries on the pull-down list, you can also type entries directly. For example, 1d 2h 5m 10s indicates a step interval of 1 day, 2 hours, 5 minutes, and 10 seconds.

 Step Forward Single-step forward by the time interval specified in **Skip Increment**

 Step Backward Single-step backward by the time interval specified in **Skip Increment**.

 Go Forward Continuously step forward by the time interval specified in **Skip Increment**. This is the control used to truly animate the display.

Go Backward Continuously step backward by the time interval specified in **Skip Increment**.

Stop Time Skip Stop **Time Skip** when using **Go Forward** or **Go Backward**.

Reset Date & Time Reset the date and time to the setting prior to the first **Time Skip** operation. After a **Time Skip** the Virtual Sky might be displaying a scene from hundreds of years in the future!

Record Trails As solar system objects (the Moon, planets, etc.) move during **Time Skip**, they leave trails behind to show the path taken. This is most effective when the time interval is no more than 10 minutes and with labels off. Also, TheSky does not animate minor planets and comets correctly when **Record Trails** is activated, so use **Tracking Setup** – described next – to hide them.

Tracking Setup This dialog box selects which solar system objects will be displayed (duplicate of same feature in the **Filters** dialog) and how trails will be recorded. The **Lock On** option allows you to keep the Sun, the Moon, or any of the planets locked in the center of the display during **Time Skip**. This is very useful to study the motion of other objects around a solar system object. The **View From** option switches the view to **3D Solar System Mode**, discussed later in this chapter.

Try this. Look north (or south if you are in the Southern Hemisphere) and select **Daytime Sky Mode** from the **View** menu or the toolbar. Set the interval to 5 minutes and click the **Go Forward** button on the Time Skip toolbar. Time really does fly!

Now click the **Tracking Setup** button and select Jupiter in the **Lock On** pull-down list. Click the **OK** button to close the **Tracking Setup** dialog box. Zoom in until you can see the names of the four bright moons of Jupiter (use *Page Down* on the keyboard or the **Zoom In** button on the toolbar). Type 10m in the **Skip Increment** box on the toolbar to set the interval to 10 minutes. Now for the show! Click the **Go Forward** button on the toolbar. Notice that TheSky clearly shows when Jupiter occults a moon as the moon clearly passes behind the planet. Moon transits are shown as the small dot of the moon passes across the face of Jupiter. If you adjusted the **Status Bar** settings as discussed earlier, you can easily see the local date and time of these events. However, TheSky Version 5 Level I does not show transits of the shadows of the moons. TheSky also accurately displays the positions of the brighter moons of Saturn.

After using Time Skip, don't forget to reset the date and time in the **Site Information** dialog box found on the **Data** menu or by clicking the **Reset Date & Time** button on the Time Skip toolbar.

Working with Sky Documents

TheSky allows you to permanently store the current settings of the Virtual Sky in a file on the disk known as a **Sky Document**. The current view of the sky, the filter settings, the date and time settings, and other vital information are all stored in the Sky Document. You work with Sky Documents just as you would a word processing file. You will find **Open**, **Save**, and **Save As** on the **File** menu. When you exit TheSky you will be automatically prompted to save the document. If you never select **File, Save As** the document in use is named **normal.sky**. TheSky automatically reopens the last document used when you restart the program. Sky Documents are saved in the User\Documents folder found in the folder where TheSky was installed on the hard drive.

Let's say you just set TheSky for the configuration you need for a star party you are attending next weekend. To save those settings, select **Save As** from the **File** menu and type an easy to remember name such as "October 10th Star Party". When you arrive at the star party and fire up your laptop, you would use **Open** on the **File** menu and select the previously saved document.

During a session of TheSky, it is likely that you will adjust filters or the date and time to suit your immediate needs. If you save when exiting, those will be the settings in use the next time you start TheSky. This is often not desirable. Generally you will have a certain configuration you find most useful. Set up this configuration and save it to a unique name like "Mike's Default Settings". To further protect this file from accidental changes make it read-only. Refer to the help system in Windows or the Mac OS for details on the read-only file attribute.

Printing Charts

Often it will not be convenient to bring a computer to your observing site. TheSky prints very nice charts to take to the field. Prior to printing, be sure to change the horizon fill to white in **Preferences** as described at the beginning of this chapter. Adjust the Virtual Sky to display the area of sky you wish to print and select **Print** from the **File** menu or click the **Print** button on the toolbar. You might first want to preview the output using **Print Preview** on the **File** menu. The area printed will not precisely match the display on the screen due to the differences in the shape of the monitor and the paper. Selecting **Landscape** in **Page Setup** on the **File** menu will allow the printed chart to more closely match the area of sky shown on the screen, but **Portrait** generally makes better use of the paper.

3D Solar System View

The 3D Solar System view is a great learning tool that clearly demonstrates the motion of the planets, comets, and asteroids in our solar system. Use an interval of 1 day on the Time Skip toolbar and click the **Go Forward** button. Try the view from directly above the plane of the solar system by clicking the **Look Up** (zenith) button. Zoom out to include the orbit of Mars. This view makes it clear why some passes of Mars are much more favorable than others. Spend some time exploring other mechanics of the solar system, such as the motions of the outer planets and comets.

Moon Phase Calendar

Found on the **Tools** menu, the **Moon Phase Calendar** is useful to plan observations of the Moon as well as nights when the Moon does not interfere with faint deep sky objects. Note that the print option makes a great wall calendar for amateur astronomers and includes the times the Sun and Moon rise in addition to the lunar phase.

Eclipse Finder

A unique tool that not only provides a list of upcoming and previous solar and lunar eclipses, but also allows you to simulate on the screen as it will appear from your location.

Select **Eclipse Finder** on the **Tools** menu. Select an eclipse from the list and click the **View** button. Use the **Go Forward** button on the Time Skip toolbar to animate the event. The image shown is the event as it will appear from the location entered in the **Site Information** dialog box on the **Data** menu.

Not all events will be visible from your location and TheSky notes when an eclipse is not visible to you. Also note that the "Show Path of Totality" option applies only to solar eclipses – your location shows on the globe as a white dot. Note that when TheSky reports that a solar eclipse is not visible from your site, you should still use the View and Time Skip – the event might be visible as a partial eclipse from your location.

Don't Stop Here

While we have covered a majority of the features of TheSky, there are more to explore. Take the time to read the User's Guide found on the Start menu and begin saving your pennies for the upgrade to Level II!

NexStar Observer List

The NexStar 60/80/114/4 models ship with a CD version of NexStar Observer List (NSOL), written by me. NSOL is also available free for download from my NexStar Resource Site (http://www.NexStarSite.com). Version 2 is on the CD and after finishing this book I will be working on version 3. By the time you are reading this, version 3 will likely be available for download.

What Is It?

NexStar Observer List is session planning software for Windows 95/98/ME/NT/2000/XP. NSOL allows you to create observer lists that are automatically saved to the hard drive of your computer. You build the lists by choosing from tens of thousands of deep sky objects and stars. The NSOL database includes 39 179 objects:

- The entire Messier catalog (110 objects).
- The entire Caldwell catalog (109 objects).
- The entire NGC catalog (7840 objects).
- The entire IC catalog (5386 objects).
- The entire Abell catalog (4076 objects).
- All 15-magnitude or brighter galaxies from the UGC catalog (6009 objects).
- 10 381 stars organized by SAO number and HD number.
- 5268 double stars organized by SAO number and HD number.
- And you can add as many additional objects as you wish!

NSOL interfaces with all models of NexStar and Tasco StarGuide telescopes, allowing direct control from your personal computer (PC). NSOL allows you to more easily GoTo objects not in the NexStar database, such as objects you learn about in magazine articles and books.

You can create as many observer lists as you want and you can exchange your lists via email with other NSOL users. Submit your favorite observer lists to the author (swanson.michael@usa.net) and they will be posted on the NexStar Resource Site (http://www.NexStarSite.com) for others to download and enjoy.

Installation

If the installation program doesn't start after inserting the CD, open **My Computer**, double-click the CD drive, and then double-click the **Setup** icon. If you downloaded from the Internet, "unzip" the file into a temporary location and double-click the **Setup** icon found in the temporary location. After installation, you may delete the temporary folder with the installation files; however, I recommend you keep a copy of the zip file.

It is important to note that if you currently have an older version of NexStar Observer List installed, you must uninstall it **prior** to installing this new version. Go to the Start Menu, Control Panel, Add/Remove Programs, and double-click NexStar Observer List. If you are prompted whether you want to remove shared components, answer **No**. This will leave your current observer lists and User Supplied Objects intact as well as other components that may be used by separate programs in your computer.

Very important: During the installation process, you may be presented with dialog boxes stating that the installation is trying to replace an already existing file on your system and that the existing file is newer. You will be asked if you want to keep the existing file: answer Yes. Many programs will present you with this option during installation and 99% of the time you should keep your existing file. Additionally, you may be presented with a dialog box stating that certain system files are out of date on your computer and need to be updated to continue. These files have been updated by Microsoft and are needed to access the database in NexStar Observer List. The updated files address security concerns that have been discovered in many Microsoft products that access databases. If you wish to use NexStar Observer List you will need to allow the setup program to update these files.

If the program won't start after installing and your computer uses Windows 95 or Windows 98, revisit the Downloads page on http://www.NexStarSite.com to download and install the DCOM patch or run the appropriate DCOM installation from the NSOL CD.

As with any Windows program, you will note that some menu items have shortcut keys listed, such as *Ctrl+N* for creating a New List. Also, menus can be accessed by Alt and the underlined letter in the menu name. The same applies for buttons in the screens you are working in; *Alt* and the underlined letter performs the same action as clicking the button with the mouse.

Initial Configuration of NexStar Observer List

When you run the program the first time, you are presented with the **Setup** dialog box. It can also be found as **Setup** on the **Tools** menu. Most of the settings should be easy to understand, but there are some potential pitfalls:

- **Scope Type** If you have a NexStar 5i/8i, select the NexStar 8/11 GPS option. If you have a NexStar 60/80/114/4 with the new GT hand control, select the "made after Dec 2001" option. If you are not certain which GT hand control you have, it is safe to try any of the options to see which works. If you have a Tasco StarGuide scope, select the option for the NexStar 60/80/114/4 made before December 2001. The "made before Dec 2001" option is only available on the download version from my web site, not the CD version.

- **Comm Port** Select the communications port the telescope is connected to. See Chapter 10 for details on connecting your computer to your telescope.

- **Time Zone Information** Generally you should select **Use Windows Time Zone Setting?** as this allows the program to calculate the correct universal time (UT) without requiring that you determine whether your location is currently observing daylight savings time or standard time. It will only work if you have set the correct time and selected the correct time zone in Windows. Double-click the time in the lower right corner of the Windows screen to confirm your settings. If your time zone is not available in Windows, you will need to remove the check mark on this option and manually enter the number of hours offset from UT.

- **Observing Location** If you do not know your longitude and latitude, I recommend the web site Heavens Above – http://www.heavens-above.com/countries.asp.

- **Altitude Controls** Prior to sending a GoTo command to the scope, NSOL will warn you if an object exceeds these limits. It is generally best to set these limits to match the Slew Limits in the telescope's hand control.

- **Location of Observer Lists** Although these will have been sent initially to the Data folder in the location where NSOL was installed, you might prefer a different location. For example, you could create a folder named "NSOL Data" in the My Documents folder. After creating the folder you must move all the files currently in the original Data folder to this new folder. Then use this option in the **Setup** dialog box to configure NSOL to use the new location. **Caution:** If NexStar Observer List cannot find the file NSObserverUserList.mdb (the User Supplied Objects database) in the folder you select, you will not be able to edit observer lists.

One final note about Observing Location and Time Zone settings. NSOL uses location and time zone along with the current Windows system time to calculate the current altitude of objects and to draw sky charts. You do not need to be too precise; the observing location and time settings have absolutely no effect on the accuracy of GoTo operations. If you are within 100–150 miles of the correct settings and a few minutes of correct time you will hardly notice the difference.

Using the Hyper Hand Controller

The Hyper Hand Controller (HHC) allows you to directly access the entire NSOL database and direct your NexStar scope to GoTo any object in the database. The HHC is also the best way to get acquainted with the methods used to access the NSOL database. You can open the HHC from the **Tools** menu or with the toolbar button resembling a NexStar hand control.

As seen in Figure 7.1, the HHC has four areas. The upper right portion of the window contains the eight separate lists that comprise the NSOL database. Use these lists to find the object you want to direct your telescope towards. You access each list by clicking the tabs along the top. Each list also has a search section that allows searching by **Object Number** or **Common/Alternate Name** (labeled simply "Name" in the column headings at the top of the lists). While you might not mind scrolling down the Messier or Caldwell lists, most of the other lists are very long and the search feature will be a big time saver. Note that HD numbers for stars and double stars are found in the Common/Alternate Name field. Simply type your search criteria in the **Search for** box, select whether you are looking for the number or name, and click the **Find First** button to locate the object in the list.

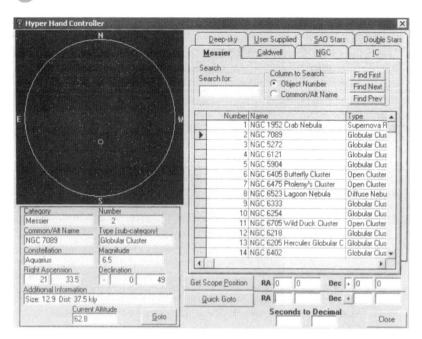

Figure 7.1. NexStar Observer List's Hyper Hand Controller window.

With most of the lists, searching in the Object Number column requires an exact match. For example, if you search for object number 10 in the Messier list, the only item found will be Messier 10; Messier 110 will not be found. The behavior is different in the Common/Alternate Name column. If you search the Double Star list for *alpha* in the Common/Alternate Name column, the first object found will be SAO 308 – Alpha Ursae Minoris. If you press **Enter** or click the **Find Next** button, NSOL will locate SAO 15384 – Alpha Ursae Majoris. If you click **Find Next** too many times and move past your quarry, click **Find Prev** to search backwards and move to the previous match.

The Messier, Caldwell, NGC, and IC lists (note that they are all on the same row of tabs) can be sorted by Object Number (the default), Constellation, Magnitude, and Right Ascension. This can be very useful to help you select items for viewing. You may find the Constellation sort order to be most useful as it allows you to easily cover all objects in a specific area of the sky. To select any of these sort orders, simply click once on the **Number, Constellation, Magnitude,** or **RA** column heading. Deep Sky, User Supplied, SAO Stars, and Double Stars can be sorted by Object Number (the default) and Right Ascension. Right Ascension can be useful when trying to match an object whose name is uncertain – especially true of double stars.

One of the eight lists is different from the rest – the User Supplied Objects. Using the Edit button to the left of this list, you can add, edit, and delete items in this list. Enter objects not found on the other lists or objects you would like to list by a different name. For example, the Double Cluster is found as object number 14 on the Caldwell list or the two clusters are known as objects 869 and 884 on the NGC list. If you find that hard to remember, you might want to add an object to the User Supplied list and simply use

List of Tables

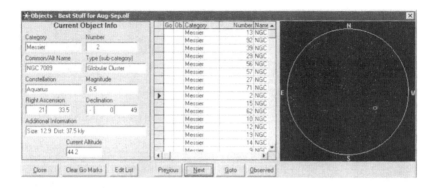

Figure 7.2. An observer list in NexStar Observer List.

The sky chart on the left shows you the location of the object you have selected from your observer list, or the object you have selected from the tables in the NSOL database. Note that it defaults not to current date and time (unless that happens to be between 1 and 6 a.m.), but rather to 10 p.m. of today's date. You can change the date and time of the sky

Figure 7.3. Editing an observer list in NexStar Observer List.

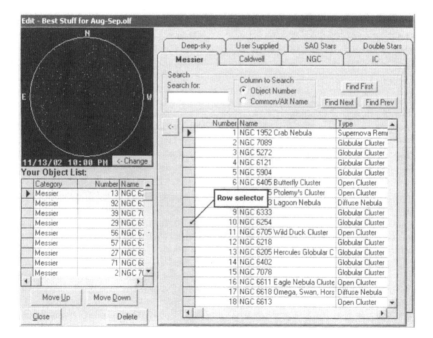

chart with the **Change** button. This sky chart makes it easier not just to organize the order of the items in your observer list but also to insure you select objects that will actually be above the horizon at your expected viewing time.

Below your object list, in the bottom left corner, there are four buttons:

- **Close** Closes the Edit List window and returns to the main window. Changes you made are automatically saved.

- **Move Up/Move Down** Use these buttons to move objects up or down the list to put things in the order in which to view them. This is the most time-consuming part of the whole process and one which will be automated in the next version. Use the sky chart to help you group objects such that you will not be jumping all over the sky as you work down through your list.

- **Delete** Removes an object from the observer list; you will be prompted to confirm the deletion.

After you have added the items you want in the observer list, click the Close button to return to the main window previously shown in Figure 7.2. Now you can see that the list window has three main sections. In the center is the list itself. To the left is full information for the object currently selected. To the right is an all-sky chart with the object currently selected shown as a small, light-blue oval. If you do not want to edit the list further and you do not wish to observe the objects with your telescope at this time, you would click the Close button. When later you want to open one of the lists you have created, use **Open Existing List** on the **File** menu or the yellow folder button on the toolbar. We will describe using an observer list to control your telescope later in this chapter.

Printing an Observing List

To print an observing list, you must first have the list open. Use **Print** on the **File** menu or the toolbar button with a printer depicted. Choose whether you want a **Compact List** (prints the objects with little space between each) or a **List with Comment Area** (prints the objects with room to write observing notes after each). You can also select the number of copies you would like printed. Even if you don't intend to use a computer to control your telescope, you might find these lists useful during your observing session. I regularly use the **List with Comment Area** to take observation notes next to my scope.

Alignment Star Chart

NSOL can also help you to locate alignment stars. Use **Alignment Star Chart** on the **Tools** menu or the button with the star-studded black circle to display an all-sky chart with NexStar alignment stars designated. This feature is dependent upon the correct date, time, and time zone in Windows as described in the Initial Configuration section above. Note that you can turn off and on the names of the alignment stars with the **Toggle Names** button.

Red Screen Mode

When you are observing at night, the normal screen colors can ruin your night vision. To change the screen colors to a scheme friendlier to dark-adapted eyes, use **Red Screen** on the

Tools menu or click the button on the toolbar with the red-lens glasses. To change back to a normal color scheme, repeat the above process, or simply close the program. In the event of this program or any other astronomy program locking up or exiting abnormally leaving you with a red screen, you can restore your normal colors by right-clicking on the desktop (the screen background when no windows are visible) and selecting **Properties**. Then go to the **Appearance** sheet and choose the **Scheme** named **Windows Standard**.

Using the Program with Your Telescope

Caution: Just as when you are using the hand control or any other program to slew your telescope, it is possible for the telescope to make contact with an immovable object – generally the tripod or mount. For the NexStar 4/5/5i/8/8i and the GPS models, this is normally not a problem unless you are using a 2-inch diagonal or perhaps a camera. For the NexStar 60/80/114, it is easy for the telescope to make contact with the tripod. For these reasons you should monitor your scope while in motion and be prepared to cancel the slew. Generally you can cancel a slew with any of the direction buttons on the NexStar hand control. Note that the **Altitude Controls** described in the "Initial Configuration" section above can help to prevent this.

Also, the **Slew Limit** settings on the NexStar 5i/8i, 8/11 GPS and the new 60/80/114/4 GT hand control will prevent NSOL and other programs from slewing to an object outside of the slew limits. If your scope does not move after sending a GoTo command, check the slew limit settings in the hand control.

The original GT hand control for the NexStar 60/80/114/4 GT has a fault that generally causes it to freeze when it receives two GoTo commands, one after another, via the RS-232 port. This is not a problem with NSOL, but rather it happens regardless of what program you use. If the hand control locks up, you will be forced to turn off and restart the scope, requiring a repeat of the scope's alignment procedure. I have found you can almost always prevent this by waiting until the GoTo is complete, then pressing any of the direction buttons – for example, to center the object in the field of view. If this doesn't work for your scope, try pressing the **Undo** button on the hand control two or three times after each GoTo has completed.

When you are ready to use NexStar Observer List to control your telescope, set up and align your telescope as you would normally. Remember the Alignment Star chart if you need help locating two alignment stars. Connect the serial port of your computer to the jack on the bottom of the hand control. If your model of NexStar has an RS-232 mode in the hand control menu, you must enter that mode for the telescope to communicate with NSOL. You can test whether your PC can communicate with your scope by clicking the **Get Scope Position** button in the NexStar Observer List main window.

 ## Using the Hyper Hand Controller to Control Your Scope

Open the Hyper Hand Controller (HHC) to directly access items in the NSOL database. To slew the telescope to an object, check to be sure it is the current object – look for the arrowhead in the row selector for that object or check the object information in the lower left of the HHC window. Notice the current location of the object as shown in the all-sky chart in the upper left of the menu. Click the **Goto** button to direct your telescope to slew to the object. You can also slew to an object by double-clicking its row selector in the list.

You can also issue a slew command for any right ascension/declination using the **Quick Goto** section in the lower right of the HHC window. Simply enter the coordinates as described earlier in the Hyper Hand Controller section of this chapter and then click the **Quick Goto** button.

After slewing to an object, you might be interested in checking the coordinates where the scope landed. Click the **Get Scope Position** button at the bottom of the Hyper Hand Controller. This readout does not update automatically; to refresh with the current scope position, click the button anytime. Note that the original GT hand control used with the NexStar 60/80/114/4 has a defect that incorrectly reports the current coordinates when the telescope is pointed to a location between 12 and 24 hours of right ascension and any negative declination.

Using an Observer List to Control Your Scope

Open the observer list(s) you wish to view. You can open up to eight lists at the same time; the **Window** menu lets you switch between all open lists. Refer to Figure 7.2 earlier in this chapter for the observer list window.

The **Next** and **Previous** buttons let you move through the list, or you can simply click on objects in the list in the center. To slew the telescope to an object, check to be sure it is the current object – look for the arrowhead in the row selector for that object or simply check the **Current Object Info** on the left. Make note of the object's location in the sky by refer-ring to the all-sky chart on the right. You might also refer to the **Current Altitude** listed on the left. Then click the **Goto** button. You can also slew to an object by double-clicking its row selector in the list.

After a slew command is issued, NSOL marks the object with a "Y" in the **Go** column. You can clear one or all of the GoTo marks with the **Clear Go Marks** button. Also note that you can mark the objects you actually observed by clicking the **Observed** button. You can clear the observed mark by clicking the **Observed** button again – it simply toggles the "Y" on and off.

Use the **Quick Goto** section at the top of the main window to slew to any right ascen-sion/declination. It works as described previously for the Hyper Hand Controller. Click the **Get Scope Position** button at the top of the main window to query the scope for its current coordinates. This readout does not update automatically; to refresh with the current scope position, click the button anytime. As noted earlier the original GT hand control used with the NexStar 60/80/114/4 has a defect that incorrectly reports the current coordinates when the telescope is pointed to a location between 12 and 24 hours of right ascension and any negative declination.

Download the Newest Version

By the time you read this, version 3 should be available for download from my web site – http://www.NexStarSite.com. Version 3 will include enhanced filtering and sorting, observa-tion log entry/printing, printing of all-sky charts, automatic list organizing, and more. Send your comments on additional features and improvements to swanson.michael@usa.net.

Chapter 8

Accessories for Your NexStar

It certainly doesn't take long before most amateur astronomers discover that their new telescope came with just enough to get them started. While you won't need every accessory introduced in this chapter, I bet you will end up with a few of them. My best advice is to take your time and only buy things that you seem to feel are lacking when you are actually out under the stars. If possible, join an astronomy club. Chances are one of the members has the item you are considering and will allow you to try it out before you buy. If this isn't possible, do some research on the Internet, in magazines, and in books. Refer to Appendix A for the web site addresses of retailers and manufacturers mentioned in this chapter.

Some of accessories are inexpensive, others will benefit from slight modifications, and others still you can make yourself. In some ways, this might be the most interactive and enjoyable chapter in the book.

Tripods and Related Accessories

A sturdy tripod is a must for steady, comfortable viewing at the eyepiece. If the tripod shakes every time you touch the telescope, focusing will be very frustrating as the image dances in the eyepiece. A shaky tripod is also tiring on a windy night. A tripod that adjusts in height allows you to set it low to the ground for maximum stability or higher for comfortable viewing when standing.

Most NexStar models come with an included tripod that is well matched to provide good stability. With some models you must select the tripod you prefer, or simply use the telescope on a stable, flat surface. You may decide to switch to a heavier tripod for added stability or a lighter tripod for easier portability. Following are details on tripods by scope model.

Tripods by Telescope Model

NexStar 60/80/114 These little NexStars come with a sturdy, adjustable-height, lightweight aluminum tripod. The tripod and fork arm mount use a unique, cup-shaped mating point that provides good stability and easy assembly, even in the dark, but

Figure 8.1. Celestron
model 93497 tripod.
Photo courtesy Celestron.

precludes the use of any other tripod. Periodically you should insure all screws and bolts are tight – loose hardware makes any tripod very wobbly.

Nexstar 4   The NexStar 4 does not come standard with a tripod. When you first start out you can simply find a sturdy table to set it on, but before long you will probably want a tripod to provide a portable, stable platform. The recommended tripod is Celestron model number 93497 (Figure 8.1). This tripod features a simple wedge incorporated into the head and stainless steel round tubular legs. The wedge allows the scope to be polar aligned for long-exposure astrophotography. The spreader arms include holes for five 1.25-inch and two 2-inch eyepieces. This tripod is sturdy, well built, and a fine match for the NexStar 4.

NexStar 5 and 5i   The NexStar 5 has at times been sold with and without a tripod as standard equipment. The NexStar 5i does not include a tripod, except in special bundles. When a tripod has been included with the NexStar 5, it was a lightweight aluminum model – 93593 – that did not always prove sturdy out of the box. In most cases all that was needed was to tighten all screws and bolts to provide good stability. In other cases it has been necessary to take more drastic steps.

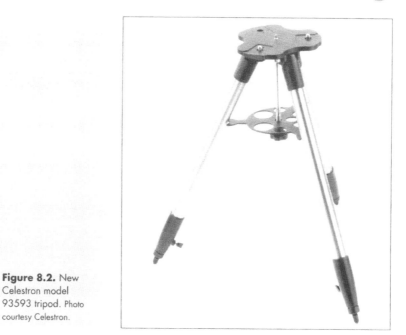

Figure 8.2. New Celestron model 93593 tripod. Photo courtesy Celestron.

If you have an aluminum tripod that remains unstable after tightening all hardware, try this. Remove the screws holding the aluminum tubing to the brackets at the head of the tripod. Visit a hardware store and pick up the next larger diameter (thicker) screw as a replacement.

For the NexStar 5 and 5i, Celestron currently recommends either the wedge-tripod (model number 93497) described above for the NexStar 4 or, for added stability, a new model number 93593 tripod. This model 93593 (Figure 8.2) is similar to the model 93457 wedge-tripod with stainless steel tubular legs, but it has no integrated wedge. The spreader arms have openings for five 1.25-inch and two 2-inch eyepieces. The added stability will be welcomed if you mount additional equipment on your NexStar 5/5i, such as a 35mm camera mounted piggyback on the scope. Since the old 93593 tripod with aluminum legs is quite inferior to the new steel leg 93593 tripod, it is best that you insure you are getting the latter before purchasing.

NexStar 8 and 8i The NexStar 8 was generally sold with the same lightweight aluminum tripod, the old model 93593 that occasionally came with the NexStar 5. The added weight of the NexStar 8 definitely stretches this tripod past its limits, so the same suggestions as described above for tightening hardware and possibly replacing screws would apply.

For economy and portability, the new steel leg model 93593 tripod is recommended. If you intend to use a wedge for polar alignment, a sturdier tripod is called for. Model 93501 (Figure 8.3), the Heavy Duty Tripod, will easily handle the load. You might note that it appears identical to the tripod supplied with the NexStar 8/11 GPS. They are similar, but the center pin on the NexStar 8/11 GPS tripod is larger and will not fit the base of the NexStar 8 and 8i.

Figure 8.3. Celestron model 93499/93501 tripod. Photo courtesy Celestron.

NexStar 8 GPS and 11 GPS
Celestron has always sold these scopes with the Heavy Duty Tripod, model 93499 (Figure 8.3). However, some distributors have separated the scope and tripod, selling them individually. The tripod is a good match for the big NexStars and offers fine stability when the scope is used in altitude–azimuth mode (no wedge). But, when mounted on a wedge, the model 93499 tripod does not offer optimum support – particularly for the NexStar 11.

Model 93499 includes three small, black, aluminum inserts for mounting a wedge on the tripod. This allows use of the three alternate holes on the head of the tripod to insure clearance for the azimuth adjuster on the Celestron Deluxe Wedge.

In 2003 Celestron introduced a new tripod, model 93509 – the Super Heavy Duty Tripod. Designed with wedge work and astrophotography in mind, this tripod should offer much improved stability. It is particularly recommended for the NexStar 11. When using a wedge, NexStar 11 owners have also used the Meade Giant Field Tripod and the Losmandy tripod. Consider the added weight and expense before settling on any of these more advanced tripods.

Another popular mounting solution for the NexStar 8/11 GPS is pier mounting. Although a few companies make portable piers, notably Astro-Physics, most piers are permanent fixtures in a yard or observatory. Many piers are at least partly custom-made, with concrete and steel being very popular material. There are about as many ways to build a pier as there are people who own them. A quick search on the Internet will turn up several detailed plans as well as vendors for pre-built piers. One common solution is a concrete pier with mounting bolts for a standard wedge. With this setup, you can still easily remove your scope from the pier to use on the tripod at a remote or alternative location.

Tripods by Model Number

Table 8.1 summarizes compatibility between NexStar telescopes and Celestron tripods. An F indicates a fit, though **not** recommended; R-W indicates recommended for use on a

Table 8.1. Compatibility between NexStar telescopes and Celestron tripods

Celestron model numbers	N4	N5/5i	N8/8i	N8GPS	N8/11GPS
93497*	R-W	R-W	F		
Old 93593†	F	R-W	F		
New 93593‡	F	R-W	R-A		
93501	F	R-W	R-W		
93499				R-W	R-A
93509				R-W	R-W

Notes:
* Includes built-in wedge.
† Aluminum legs.
‡ Steel legs.

wedge or in altitude–azimuth mode; R-A indicates recommended for altitude–azimuth mode only.

Tripod Modifications and Accessories

It seems that the perfect tripod has yet to be designed. Or, more correctly stated, every telescope owner seems to have a different idea of what the perfect tripod would be. Perhaps we feel it isn't stable enough or it is too heavy for easy portability or there are not enough places to hold the various items we want immediately accessible. This situation is definitely a case of "necessity is the mother of invention," with many ingenious improvements created by amateur astronomers around the world.

In addition to tightening screws and bolts and potentially replacing screws with the next larger size, some have used nylon washers between loose-fitting hardware to remove any additional play. For example, the bolts attaching the accessory tray arms to the tripod legs on the NexStar 60/80/114 have substantial side-to-side play and benefit from this suggestion. Another common stability improvement is to hang a heavy weight – the battery pack powering the scope is most common – from a hook-bolt added to the bottom of the accessory tray arms. Others have gone so far as to fill the legs with lead shot or other heavy material to improve the tripod's ability to dampen vibrations.

Joe Shuster developed an inexpensive and very effective enhancement for the original aluminum NexStar 5/8 tripod, model 93593. As seen in Figure 8.4, this tripod has a threaded hole in the center of the bottom of head. A hexagonal spreader can be easily fashioned from a small piece of wood, with a bolt and wingnuts allowing hand tightening to greatly improve stability.

The NexStar 8 GPS and 11 GPS include Celestron's Vibration Suppression Pads to improve tripod stability. The pads are made with a plastic cup and circular disk with viscous gel-like foam sandwiched between. A pad is placed under each tripod leg to help dampen vibrations (Figure 8.5). They definitely work as advertised and are highly recommended for the NexStar 5 and larger. The smaller NexStars do not benefit as much due to their lighter weight. In addition to Celestron, Meade and Orion also make vibration pads. Orion's are the least expensive, with the added benefit of little reflective spots on top of the pad to help make them visible when you are sweeping the area after packing up for the night. Many a set of vibration suppression pads has been left on the field after a long observing session. One final suggestion – get a small plastic food container with lid to store them in. The pads often pick up mud and grass and the container will let you transport them more easily.

Figure 8.4. homemade spreader for original 93593 tripod. Photo courtesy Joe Shuster.

Setting the NexStar 8/11 GPS on the top of the tripod and aligning it with the pin has proven difficult for many. After a little practice you will likely find it easier than your first nervous attempt. If you continue to find this difficult, Starizona created the Landing Pad just for you. The Landing Pad is mounted on top of the tripod and easily guides the scope onto the center pin. Additionally, it includes openings for the small ovals on the bottom of the scope base to help line up the boltholes. For less than $40, many will find this a great accessory.

The three bolts provided to mount your scope to the tripod work OK in the Alt–az mode, but when using your NexStar GPS on a wedge for astrophotography, it is necessary to tighten them more than can be accomplished by hand and undoing them can be nearly impossible. You can use a hex wrench with the existing bolts, or you might want to get a set of Ray (RJ) Rauen's steel replacement bolts. At $65, they feature larger heads to allow easier hand tightening. RJ also has $20 bolts with large plastic knobs for more budget-minded individuals. Both are available at http://www.hometown.aol.com/rjrjr001. Morrow Technical Services (of Bob's Knobs fame) also offers plastic knob tripod bolts for $29. Visit their web site at http://www.bobsknobs.com for details.

The accessory tray for the NexStar 60/80/114 tripod is quite small, but this is easily remedied. The simplest improvement is a 10-inch cake pan. Using the accessory tray as a template, mark the position of the three bolts and drill $\frac{1}{8}$-inch holes. Set the original tray inside the cake pan, set the assembly on the tripod arms, and then use the thumbscrews as normal to fasten it all together. A little more work can result in even more utility. As shown in Figure 8.6, a plastic serving tray was drilled for eyepieces and slots were cut for holding the hand control in three different locations in this tray by Greg Childress.

If you have the Heavy Duty Tripod with the simple leg support bracket (spreader), you are cursed with no accessory tray at all. Star Tek Industries (http://www.startekind.com) manufactures a replacement for the spreader that doubles as an accessory tray. If their

Figure 8.5. Vibration
suppression pad in use.

Figure 8.6.
Homemade accessory
tray for little NexStars.
Photo courtesy Greg
Childress.

Figure 8.7.
Homemade eyepiece
tray for Heavy Duty
Tripod.

3-in-1 tray is too expensive for you, you can easily make an accessory tray for just a few dollars and a few hours of your time.

I made the tray in Figure 8.7 from $\frac{1}{4}$-inch plywood. Rather than replacing the support bracket, it simply mounts beneath the existing bracket. Using a jigsaw or bandsaw, cut a 16-inch (40-cm) circle from the plywood. Next drill a $\frac{1}{2}$-inch hole in the center for the threaded rod. Then use $1\frac{1}{4}$-inch and 2-inch hole saws to make eyepiece holes as desired. The tray pictured here has sixteen $1\frac{1}{4}$-inch eyepiece holes and leaves the third section untouched to provide a flat surface for other objects. It is a good idea to then sand all the edges and surfaces.

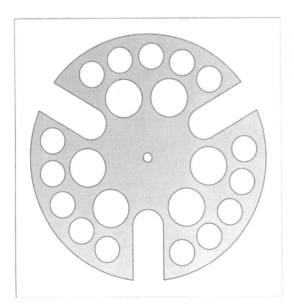

Figure 8.8. Design
for Heavy Duty Tripod
tray.

You may want to paint the tray (helps to protect the wood and your eyepieces) and you may also want to cut three 2-inch-wide slots in the tray as shown in Figure 8.8 to allow the legs to fold without removal of the tray.

Wedges

While NexStar telescopes can track the sky when mounted directly on the tripod in altitude–azimuth mode, astrophotography with exposures longer than about one minute will suffer from star trails due to rotation of the field of view. Longer exposures require the scope to be polar-aligned. For NexStar scopes that means using a wedge. We will discuss more details regarding astrophotography in Chapter 11.

Tripod model 93497 has a built-in wedge, but for all other tripods you must purchase a wedge separately. There are no wedges made for the NexStar 60/80/114 tripod and their base design does not readily allow them to be mounted on alternate tripods. In any case, the tracking on these models is very coarse, so polar alignment usually does not prove practical unless you are ready to manually guide long exposures. Nonetheless, a few adventuresome souls have actually built wooden wedges or tripod adapters for the little NexStars.

For the NexStar 4, tripod 93497 will give good results. For the NexStar 5/5i, tripod 93497 will get you started; however, if you intend to use a 35mm camera, a piggy-backed guide scope, or other heavy accessories, you will want to use tripod 93593 with a wedge.

Celestron makes two wedges: the NexStar Wedge (model 93658) and the Heavy Duty Wedge (model 93655; Figure 8.9). Mounted on the new model 93593 tripod, the NexStar Wedge is a good match for the NexStar 5/5i with heavier accessories. Mounted on tripod 93501, the NexStar Wedge provides good support for the NexStar 8/8i.

The Heavy Duty Wedge is required for the NexStar 8 GPS and 11 GPS. As mentioned earlier in the section on tripods, the stock tripod (model 93499) is adequate for the NexStar 8 GPS on a wedge, but not really stable when using the NexStar 11 GPS on a wedge. For the 11 you will need to use one of the larger tripods previously discussed. Also, there is an upgrade kit available for the Heavy Duty Wedge to allow easier adjustments in latitude and azimuth. This is not really an option; adjusting the wedge without this kit is nearly impossible.

Ray (RJ) Rauen offers a complete set of 10 no-tools bolts for the Celestron heavy-duty wedge. At just $53, including shipping, they are an absolute necessity for field use of the wedge. See them at RJ's web site: http://www.hometown.aol.com/rjrjr001.

A few companies build wedges with enhanced stability and precision latitude and azimuth adjustment mechanisms. They are generally quite expensive and they often have a waiting list. But if you are looking for added stability, there are options available. Consider Superior Wedge from Mitty Observatory Machine Shop (http://home.cfl.rr.com /mitty/Superior_Wedge.htm), the Mettler Wedge (http://www.greatnorthern.net/ ~jam/links/Manuf/mettler.htm), the Milburn wedge from Bonney Lake Astro Works (http://www.milburnwedge.com), or Le Sueur Manufacturing's Aurora Polaris Pier and Wedge (http://www.astropier.com). Before purchasing a wedge, make sure it will be compatible with your intended tripod or pier and the base of your scope.

Power Sources

It won't take long to realize that using AA batteries is an expensive way to feed your NexStar. The NexStar 8/11 GPS model does not even provide that battery option. AA bat-

Figure 8.9. NexStar 11 GPS on Celestron Heavy Duty Wedge. Photo courtesy Dave Cole.

teries can be useful for emergency power, but not general operation. One of the first accessories you will need for your telescope is a reliable power source.

Even brief interruptions in power will cause problems with a computerized telescope. Erratic behavior or a complete loss of alignment is likely to result. The power connector on NexStar telescopes is often the source of such problems. It is not the common size for such connectors and, as such, many a.c. adapters and power cables do not result in a tight fit. The correct connector for NexStar scopes has an outside diameter of 5.5 mm and an inside diameter of 2.1 mm. While a loose-fitting connector might seem to work, hard-to-diagnose problems often occur.

If you are near a power outlet, an a.c. adapter is a good option. Almost any adapter will do as long as it provides 12 volts d.c., at least 1.5 amps (1500 milliamps), and a connector wired with the inside positive (+) and the outside negative (−). Most stores stock such adapters, but I recommend you purchase a higher-quality adapter typically found in an electronics store even if it costs a little more.

When you are not near a power outlet, what you really want is a rechargeable power source. In fact, I always use one of these. They provide more reliable power than the typical a.c. adapter and there are fewer cords to deal with. NiCad and NiMH batteries won't do (unless you make your own battery holder), as they only produce 1.2 volts as compared with the 1.5 volts of an alkaline battery. NexStar telescopes all require 12 volts to run correctly, so you would need 10 correctly connected NiCad or NiMH batteries for reliable operation.

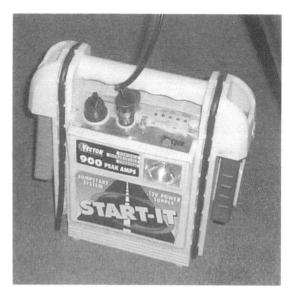

Figure 8.10.
Automotive jump-start
battery.

12-volt rechargeable power sources are available from many astronomy vendors – sometimes they are referred to as field power packs. Such power sources can be found in the camping and outdoor equipment sections of many stores. Most automotive stores carry larger, higher-capacity power sources sold as portable automotive jump-start systems (Figure 8.10). For the ultimate in power, you can construct a power source using a marine deep-cycle battery. Almost all of these power sources use a type of battery known as a gel cell. These batteries are very similar to the lead–acid batteries used in automobiles. Be aware that these batteries are usually considered hazardous items and shipping charges can be very high. For that reason it is usually best to purchase locally.

With any of these options, the requirements you are looking for are 12 volts, 7 amp hour or greater (this is a measure of how long the battery lasts), a "cigarette lighter" socket, and a recharger. All but the smallest rechargeable 12-volt power packs are 7-amp-hour or more. A 7-amp-hour battery will run a NexStar scope for an entire night, but this will stress the battery. 15 amp hours is the minimum recommended for all night sessions. If you use a dew prevention strip, a second 15-amp-hour battery just for this purpose is recommended. Laptop computers, cameras, and other devices will also increase your battery capacity requirements.

For most 12-volt power packs, you will need to supply the correct cord to go between the power pack and the scope. Most astronomy shops that sell Celestron equipment can provide the correct cord or you can try an electronic parts store. The cable you are looking for has a cigarette lighter plug on one end and the correct 5.5mm/2.1mm plug as described earlier. Again, be very careful that positive (+) is on the inside conductor.

A few additional notes on gel cell power sources. To prolong the life of the battery, recharge a gel cell after every use. Do not wait until the telescope starts acting sluggish or the meter on the battery indicates low power. Also, overcharging can easily damage a gel cell and most do not include a recharger with automatic shutoff. NexStar owner Phil Chambers provided the following tip. Plug the charger into a power outlet timer (the little gizmo used to turn lights off and on to deter burglars while you are away) with all the "ON" pins removed and only a single "OFF" pin inserted. Then you can set the timer to shut off after the recommended recharge period. Another option is to purchase a trickle charger from a marine or automotive parts store. While more expensive, it will keep your battery in top condition. Be aware, though, that some gel cell power sources have electronics that prevent a trickle charger from working properly, something you will most likely determine only through trial and error.

Eyepieces and Related Items

The adage goes that the eyepiece is half of your telescope, so don't settle for anything but the best. Indeed this is largely true. The finest telescope will provide terrible views through an inferior eyepiece. But it is also true that as the price increases beyond a certain point, only marginal gains in quality are realized. Nonetheless, the eyepieces provided with the NexStar scopes truly are basic level eyepieces and much improvement can be had with better models.

The best general advice I can give you regarding eyepieces is to try before you buy and take your time. An eyepiece that looks perfect on paper may not perform up to your expectations. In fact, every eyepiece performs differently in different telescopes. For example, a long-focal-length Plössl might work well in the long focal length of the NexStar 11, but provide distorted views in the shorter-focal-length NexStar 114. Additionally, variations in eye relief and field of view make eyepiece selection a very personal decision. In a given scope, what works for one person may not for another. Try to get out to a star party or other astronomy club gathering to borrow eyepieces to test in your telescope. Most amateur astronomers are more than happy to help out their fellow observers. If this is not practical, purchase your eyepieces from a company with a lenient return policy.

To narrow down the search, use the Internet. Read reviews at the sites listed in Appendix A or ask questions on a discussion group. These opinions will help you to get started. And naturally I have some information and opinions too.

Eyepiece Basics

Besides basic optical quality – largely immeasurable, but certainly discernible – there are three vital statistics for any eyepiece: focal length, apparent field of view (AFOV), and eye relief. As discussed in Chapter 2, shorter-focal-length eyepieces result in higher magnifications, a larger AFOV allows more of the sky to be seen in the eyepiece (when comparing two eyepieces with the same focal length), and eye relief is the distance the eye may be from the eyepiece and still see the entire field of view (FOV).

Various eyepiece designs are available with each having slightly different characteristics. Table 8.2 summarizes the most popular designs.

Eyepieces come in two barrel sizes, 1.25-inch and 2-inch. A 2-inch focuser tube or diagonal can accept 1.25-inch eyepiece using an adapter. 2-inch eyepieces are used with longer focal lengths to provide a wider field of view. You should also note that many department store or "junk" telescopes use an older 0.965-inch eyepiece size that should be avoided.

Many eyepiece lines are parfocal – that is to say that when changing from one eyepiece to another in the series, only a slight touch up is required to achieve focus. The added convenience is welcome when switching eyepieces to find the best magnification for a particular celestial object.

Untreated lenses reflect up to 10% of the light passing through them. Good-quality eyepieces are "multicoated". This indicates that at least one lens surface was treated with

Table 8.2. The most popular eyepiece designs

Kellner	Commonly referred to as SMA – super modified achromatic. Included with intro scopes such as the little NexStars. Fairly good performance in longer focal lengths. Kellner eyepieces exhibit chromatic aberration and poor performance at the edge of the FOV. Shorter focal lengths have uncomfortably short eye relief. With the cost and much better performance of budget Plössls, there is very little reason to purchase new Kellner eyepieces, although University Optics and Orion (Explorer series) do sell better-quality SMAs for those on a very tight budget.	AFOV: 45–50°
Orthoscopic	Orthos provide very sharp, high-contrast views in the center of the field of view. Orthos provide good eye relief at short focal lengths, but they exhibit a narrow field of view. They also have relatively poor performance at the edge of the FOV. This combination of characteristics makes them a poor choice for most deep sky objects, but many amateur astronomers feel they are the best eyepieces for viewing planets. University Optics and Brandon make fine orthoscopic eyepieces.	AFOV: 40–45°
Plössl	Providing very good performance to the edge of the FOV with few optical flaws, Plössls are considered by most to be the best all-around performer. They provide a moderate FOV, but have short eye relief at shorter focal lengths. Many manufacturers offer Plössls and modified or "super" Plössls. Celestron's Ultima, Meade's Series 4000 Super Plössls, and Tele Vue's Plössls are the top contenders in this category. As is often the case, you get what you pay for, and these more expensive models do generally perform better.	AFOV: 50°
Wide-angle and super-wide-angle	Various wide-angle and super-wide-angle designs have been created for viewing deep-sky objects. These are the most expensive class of eyepiece, prized for a wide field of view at higher magnifications. They are offered under many trade names from many manufacturers. Compare the Panoptic and Nagler lines from Tele Vue, the Super Wide Angle and Ultra Wide Angle series from Meade, the Axiom models from Celestron, the Konigs from University Optics, and the XL series from Pentax. With these eyepieces you should definitely try before you buy. Watch for poor sharpness at the edge of the field of view, difficulty positioning the eye and the "kidney bean" (floating black spots) effect. Some are very heavy and will produce balancing problems with little NexStars.	AFOV WA – 60–70° SWA – 70–85°
Long-eye-relief designs	Observers requiring eyeglasses when viewing need eyepieces with longer eye relief. These designs typically offer 20 mm of eye relief. Several popular makes are on the market, such as the LV series from Vixen and Celestron, the Lanthanum and ED-2 series from Orion, the Radians from TeleVue, and XL series from Pentax.	AFOV varies

Table 8.2. The most popular eyepiece designs (continued)

	Again, it is important you test these before committing to a purchase, some users complain of poor optical performance and difficulty positioning their eye with various models.	
Zoom eyepieces	Definitely not for everyone; some swear by them but most swear at them. Images often suffer from distortion at the edge and the field of view is very small, even at longer 40° focal lengths. The only notable exception is the Tele Vue 3–6mm Zoom – but this extremely short focal length is really only a match for short-focal-length scopes like the NexStar 80.	AFOV is quite narrow, generally around 40°

multiple layers of anti-reflective coatings. Better-quality eyepieces are "fully multicoated" – an indication that all air-to-glass surfaces are multicoated.

Good-quality eyepieces are also mechanically sound and obviously well built. Eyepieces with uneven paint or coatings should be avoided. Check the inside of the barrel for a well-darkened and even appearance. Eyepieces with poor blackening inside the barrel usually result in ghost images around bright objects. A nice additional touch is a setscrew groove that insures your eyepiece stays put even if things get upside down.

Eyepieces and the NexStar Telescopes

In addition to the general information given above, when selecting eyepieces for your NexStar you should consider the maximum magnification and maximum true field of view (TFOV) for your scope.

Let's review the discussion on eyepiece calculations presented in Chapter 2. First calculate the magnification of the eyepiece:

magnification = scope focal length ÷ eyepiece focal length

Then you can directly calculate the true field of view:

true field of view = eyepiece apparent field of view ÷ magnification

Let's say the focal length of our scope is 1000 mm and the focal length of our Plössl eyepiece (50° AFOV) is 10 mm:

$$MAG = 1000 \div 10$$
$$MAG = 100 \times$$
$$\text{and TFOV} = 50 \div 100$$
$$TFOV = 0.5°$$

Obviously calculating this for every eyepiece on the market would be time-consuming and tedious. You can download an Excel spreadsheet at the NexStar Resource Site (http://www.NexStarSite.com) or visit Peter Enzerink's web site and try out his "Scopulator" (http://enzerink.net/peter/astronomy/scopulator.htm). Both allow you to enter the focal length of your telescope and the focal length and AFOV for multiple eyepieces to provide a complete table of magnifications and TFOV for each eyepiece.

Table 8.3. Focal length of eyepiece required for theoretical maximum magnification for each NexStar model

NexStar model	Maximum theoretical magnification	Focal length (mm) of required eyepiece	Maximum usable magnification*	Focal length (mm) of required eyepiece
60	120	5.8	120	5.8
80	160	2.5	120	3.3
114	225	4.4	225	4.4
4	200	6.6	200	6.6
5/5i	250	5	250	5
8/8i/8GPS	400	5	300	6.8
11GPS	550	5	300	9.3

* Usable magnification is limited either by optical design, such as the extremely short focal length of the NexStar 80, or seeing conditions. It is not an absolute value and will vary from scope to scope and night to night. On very good nights, 300× is generally the highest magnification possible. Very rarely, seeing conditions will be nearly perfect and higher magnifications will be possible.

Determining the focal length of the eyepiece required for the theoretical maximum magnification for each NexStar model is very straightforward. Table 8.3 summarizes the information for the entire NexStar line. Keep in mind that you can double the focal length of the eyepiece and still achieve the same magnification by using a Barlow lens. For example, rather than using a 5mm eyepiece, you can substitute a 10mm eyepiece when using a 2× Barlow.

Choosing an eyepiece for maximum true field of view (TFOV) is a little more complicated due to the relationship between the magnification (in other words, the focal length of the eyepiece) and the apparent field of view (AFOV) of the eyepiece. Maximum TFOV is limited by the narrowest opening in the optical path, known as the field stop, generally about 1.2 inches with 1.25-inch eyepieces and just less than 2 inches for the larger-sized eyepieces. Additionally, the 5- and 8-inch NexStars are actually limited to a field stop of about 1.5 inches, even when using 2-inch eyepieces. We will discuss 2-inch accessories in more detail later in this chapter.

Table 8.4 lists the approximate widest true field of view for each NexStar when using 1.25-inch eyepieces, 2-inch eyepieces, and 1.25-inch eyepieces with an f/6.3 reducer/corrector (discussed later).

Table 8.4. Approximate widest true field of view for each NexStar when using 1.25-inch eyepieces, 2-inch eyepieces, and 1.25-inch eyepieces with an f/6.3 reducer/corrector

NexStar model	Maximum TFOV with 1.25in eyepiece	Maximum TFOV with 2in eyepiece	Maximum TFOV with 1.25in eyepiece and f/6.3 R/C
60	2.5°	n/a	n/a
80	4.4°	n/a	n/a
114	1.75°	n/a	n/a
4	1.3°	n/a	n/a
5/5i	1.4°	1.75°	2.2°
8/8i/8GPS	0.9°	1°	1.3°
11GPS	0.6°	1°	1°

After performing all the calculations you will find that a 35mm Plössl with the usual 50° AFOV yields the maximum possible field in a 1.25in eyepiece format. This is true even when using the f/6.3 reducer/corrector. For higher magnification, a 25mm 70° AFOV wide-angle eyepiece yields about the same true field of view.

With 2in eyepieces the 5- and 8-inch NexStars provide the maximum TFOV with a 40mm Plössl or 30mm wide-angle (70°) eyepiece, due to their internal 1.5in field stop. The NexStar 11 reaches maximum TFOV with a 55mm Plössl or 40mm wide-angle (70°) eyepiece.

Barlow Lenses

A Barlow lens effectively stretches the focal length of any scope. This results in higher magnification from any given eyepiece; thus Barlow lenses are generally designated by their magnification factor. The most common Barlow doubles the magnification and is thus a 2× Barlow lens. Like eyepieces, better Barlow lenses sport multicoated optics, precise mechanical tolerances, and effective internal blackening. The typical Barlow has two lens elements, but the best include a third lens for improved edge correction and reduced chromatic (color) aberrations. These three-element or apochromatic Barlows are much better performers than their lower-quality siblings and well worth the money. Celestron's Ultima, Orion's Shorty-Plus, and the University Optics 2.8× Klee are among the best in this category.

Focal Reducer/Corrector

Reducing the focal length of the scope itself by using a focal reducer/corrector can increase the maximum field of view for an SCT. Originally designed to produce flatter fields for photographic work, the f/6.3 reducer/corrector is now used by many for visual observations. Celestron and Meade offer an f/6.3 reducer/corrector that changes the scope from an f/10 instrument to f/6.3. In other words, the focal length of the scope is reduced to 63% of normal. This is a very effective means of producing wider views in the 5in and larger NexStars. The reducer/corrector attaches between the rear cell of the scope and the visual back that holds the diagonal.

Recently introduced is a new focal reducer/corrector from Baader Planetarium. Known as the Alan Gee Telecompressor Mark II, it converts an f/10 SCT to f/5.9. It is effective for both visual and photographic work. The Alan Gee is available in Europe from Baader Planetarium and in North America from Alpine Astronomical.

Higher-ratio reducer/correctors are available for photographic work only and are discussed in Chapter 11.

Binocular Viewers

Although we all get use to it, almost everyone finds observing tiny, faint objects with one eye to be wholly unnatural. On the other hand, the view through a pair of binoculars is comfortable and quite natural. Binocular viewers are available for telescopes to give that same enhanced view.

Baader Planetarium, BW-Optic, Celestron, Denkmeier Optical (Figure 8.11), LOMO, and Tele Vue all make bino viewers. Some work better than others when used with various

Figure 8.11.
Denkmeier bino
viewers in a NexStar
11 GPS and Celestron
bino viewers in a
piggybacked Tele Vue
Pronto. Photo courtesy Jim
Gutman.

types of scopes and it is recommended you research carefully before you invest in one of these rather expensive accessories. An excellent review of bino viewers by Jim Gutman is available at the Cloudy Nights web site on page http://www.cloudynights.com/acces-sories2/binoviewers.htm. Keep in mind you will also need **two** matching eyepieces at each desired focal length. Despite these cautions, I will state that the view of many objects with bino viewers in an 8- or 11-inch telescope is simply unrivalled – well, except in an even bigger aperture!

Diagonals

Schmidt–Cassegrain (SCT), Maksutov, and refractor telescopes use a diagonal to posi-tion the eyepiece at a comfortable angle when viewing objects high overhead. The light path is reflected at a 90° angle by either a mirror or a glass prism in the diagonal. Every NexStar except the 114, a Newtonian scope, comes with a diagonal. The NexStar 60 and 80 come with a budget-grade mirror diagonal. The NexStar 4 uses an internal flip mirror that functions as the diagonal. All the SCT models, the NexStar 5 and up, come with a prism diagonal. It is possible to replace the diagonal on the NexStar 60, the 80, and the SCT models.

So, the question is, why would you want to replace your diagonal? In most cases the diagonal that came with your NexStar will last for many years and should not need to be replaced, but there are some things to consider.

Diagonal Basics

Eventually, a standard mirror diagonal will tarnish, reducing the amount of light that reaches your eye. Additionally, cleaning easily scratches the coatings on the mirror of a lower-priced mirror diagonal. This causes light to be scattered, reducing the contrast of the image you are viewing. Mirrors with dielectric coatings have a very high light throughput and their surfaces do not tarnish and should last a lifetime of cleaning. In fact, one manufacturer demonstrates how the coatings stand up when scrubbed with steel wool! (Don't try this at home, boys and girls!)

Also, budget mirror diagonals are often out of collimation – that is to say, they alter the path of light away from the center of the eyepiece. It is sometimes possible to put small shims (black paper works well) under one or two of the corners holding the mirror to correct this problem.

Prism diagonals are generally well collimated, but cleaning will take its toll on the coatings on the surface of the glass. Also, prism diagonals do not transmit as much light as a good-quality mirror diagonal. Note that most observers can hardly tell the difference in light throughput when comparing a prism with a mirror diagonal, except under the best seeing conditions.

Budget-quality mirror and prism diagonals often suffer from rough surfaces that result in optical distortion, while premium diagonals boast $\frac{1}{4}$-wave (quite good) or better performance.

The easiest way to test the quality of your diagonal is to compare the views with it and without it. The eyepiece can be placed directly in the focuser or visual back. Wait for a night with clear, steady seeing and compare by viewing several different types of objects. If there is very little difference, then the optical quality of your diagonal is quite good and replacing it will bring little improvement.

There are also mechanical considerations. The chrome barrel on the stock diagonals are threaded into the body of the diagonal. In some cases it doesn't seem to be possible to get this connection tight enough to prevent the diagonal from turning in the focuser or visual back when using heavier eyepieces. A little thread lock can help though. Higher-quality diagonals are well built and usually do not suffer from this problem.

Many replacement diagonals use a brass compression ring to hold the eyepiece in place rather than having the thumbscrews directly biting into the barrel of the eyepiece. The compression ring holds the eyepiece more securely and does not mar the barrel.

If you intend to purchase any 2-inch eyepieces, you will be forced to upgrade to a new 2-inch diagonal. The main reason for using 2-inch eyepieces is to increase the maximum available true field of view as discussed in the eyepiece section of this chapter. It should be noted that a more economical way to increase the field of view is to use a focal reducer with less expensive 1.25-inch accessories. However, there is a disadvantage when you want to go to higher magnifications, as you must then remove the focal reducer, not just change the eyepiece. Also, some prefer the view provided by 2-inch eyepieces when compared with similar fields of view provided by a focal reducer and 1.25in eyepieces.

Replacement Diagonals

Many companies make fine diagonals. If you are looking for a simple replacement for a lost or broken 1.25in diagonal, consider the basic mirror or prism diagonals from Apogee and Orion. For a basic 2in mirror diagonal for SCT scopes, Orion and Celestron both make models that screw directly to the back of the scope without using the visual back. For a well-finished, mechanically superior diagonal with a standard mirror, look into the diagonals from Tele Vue and William Optics. There are both 1.25in and 2in models to choose from.

For a dielectric or enhanced mirror diagonal, consider the Astro-Physics MaxBrite, the Tele Vue EverBrite, and the Baader Planetarium MaxBright. The Baader MaxBright is my personal choice, but it is more expensive and at the time of this writing available only in Europe and possibly by request at Alpine Astronomical (http://www.alpineastro.com). It can be used with a 2in visual back on an SCT or in a refractor telescope, just like the stock diagonal. Alternately you can remove the chrome barrel and screw it directly to the back of any SCT scope. All dielectric diagonals should outlast your telescope; they include compression rings for securing the eyepiece, and they boast light throughput above 95%.

For a 2in diagonal to be used on the NexStar 5/5i/8/8i it is recommended that you consider a model that screws directly to the rear of the scope after removing the 1.25in visual back (Figure 8.12). If you prefer a refractor-style diagonal with a barrel that inserts into the visual back, you will require a 2in visual back – most manufacturers simply call it a 2in SCT adapter – that replaces the standard 1.25in visual back. The disadvantage of the 2in visual back is the added extension off the rear of the scope. With the NexStar 5/5i, this can cause difficult balance problems. With the NexStar 8/8i the diagonal will contact the base preventing the scope from pointing towards the zenith. The Orion screw-on diagonal is not useful in this fashion as it still extends too far for clearance; rather the Celestron and Baader MaxBright are recommended. Alternatively, you can install Ray's Bracket on a NexStar 5/5i/8/8i (see Chapter 13) and have ample clearance for any type of diagonal.

Figure 8.12. 2-inch diagonal screwed directly to rear of NexStar 8i. Photo courtesy Baader Planetarium.

There are more options when using a 2in diagonal with the NexStar 8/11 GPS; both models have ample clearance when pointed to the zenith. For these scopes you will simply need a 2in visual back or adapter. Astro-Physics, Tele Vue, and Baader all make suitable adapters. The adapter for 8in SCT scopes threads directly to the rear of the NexStar 8 GPS or to the large adapter ring on the rear of the NexStar 11 GPS. The opening on the rear of the NexStar 8 GPS and the large adapter ring is just 1.5in rather than 2in. This cannot be modified on the 8in scope, but on the NexStar 11 you can purchase adapters that thread directly to the rear of the scope after removing the large adapter ring. Astro-Physics and Baader both make such adapters. Also consider the Virtual View from Starizona and EyeOpener from Peterson Engineering.

And finally, while on the topic of 2in diagonals, it is possible to use a 2in diagonal in the NexStar 80. The Telescope Warehouse sells a replacement 2in focuser unit that easily replaces the stock focuser. Simply remove the three screws holding the focuser in place and replace with the new 2in unit. While a wider field of view is not usually necessary with the NexStar 80, it does allow you to use 2in accessories interchangeably between this scope and another. Also, a 35mm film frame is vignetted with the 1.25in focuser.

Filters

Many new amateur astronomers are lured into buying filters and are sorely disappointed by their performance. Filters can be very useful, once you understand how they work and their inherent limitations. Refer to Appendix A for the web site addresses of the many companies mentioned in this section.

Filter Basics

There are five common types of filters:

- **Lunar Filter** A neutral-density (uncolored) filter to reduce the extreme brightness of the Moon.
- **Color Filter** Available in many different colors, used to enhance visual observations of the planets, and used to make color photographs with black-and-white cameras.
- **Solar Filter** Allows safe observation of the Sun.
- **Deep Sky Filter** Reduces unwanted light to aid in seeing fainter objects.
- **Minus Violet Filter** Reduces the amount of false color in short-focal-length telescopes like the NexStar 80.

All filters are designed to block light. This inherently darkens the image, so the scope must be able to pull in enough light to still allow you to see the object you are interested in. Due to this fact, small telescopes often do not benefit from colored and light pollution filters due to their lesser light-gathering ability.

Filters are available in three common sizes: 1.25in, 2in, and SCT. 1.25in and 2in filters are threaded for attachment to the bottom of most eyepieces or the front of a standard diagonal. Naturally, the eyepiece/diagonal must be the same size as the filter. SCT filters are designed to thread onto the rear of a Schmidt–Cassegrain telescope, with the visual back threaded onto the filter. With a filter attached to the rear of an SCT or the front of a diagonal you can change eyepieces and leave the filter in place.

Lunar Filters

The Moon is very bright, especially at lower magnifications. This makes it difficult to see fine detail, causes loss of dark-adaptation, and can even be a little painful. A standard lunar filter may block 80% or more of all visible light. A polarizing lunar filter uses two polarized elements that can be rotated to vary the amount of light blocked. The Moon always looks better through a filter in any size of telescope. Orion, Celestron, Meade, and others make good-quality lunar filters.

Color Filters

Color filters are mostly used for viewing the planets. By blocking certain wavelengths (colors) of light, they help to bring out faint details on the planet surface. Each color has a different effect on each planet. In particular you will find color filters useful for Jupiter, Saturn, Venus, and Mars. Other than for Jupiter and Venus (two very bright objects), color filters will not provide much benefit for scopes smaller than 4.5in. Look for quality color filters from Celestron, Baader Planetarium, Orion, and others.

Color filters are also used to produce color images with black-and-white CCD cameras. The subject is imaged through three filters, usually red, green, and blue, and the separate images are combined with computer software to create the final, color image.

Solar Filters

The Sun can be viewed directly with the proper filter. Safe solar filters attach to the front of the telescope and completely cover the opening. Never use a solar filter that attaches to the eyepiece! The intense heat concentrated at the eyepiece will cause it to crack, allowing unfiltered sunlight to flood into the eye causing permanent damage.

Conventional solar filters come in two varieties: Mylar film and glass. These filters allow us to see sunspots and granulation on the surface of the Sun. With the exception of Baader AstroSolar Film (Mylar), glass filters generally give a higher-resolution view. However, Mylar filters will not break if bumped or dropped. Glass filters are available from Orion, Thousand Oaks, and others. Mylar filters are available from many, but the clear standout is Baader Solar Film. Most consider it to be second to none, even compared with glass filters. Baader AstroSolar Film can be purchased in sheets or in pre-built filters for specific telescopes. Sheets are available from Kendrick Astro Instruments or Astro-Physics along with simple instructions for making your own filter cell (Figure 8.13). Celestron makes pre-built filters for the NexStars that can be purchased from SightAndSoundShop, Adorama, and Astronomics. Kendrick also sells pre-built filters made with Baader film.

Another type of solar filter is the H-Alpha variety. They are very expensive, but allow us to view flares and other features in the Sun's chromosphere. These filters are not suitable for most NexStar telescopes, but the SolarMax series from Coronado can be successfully adapted for use on the NexStar 60 and 80.

Deep Sky Filters

Deep sky filters are designed to block unwanted light while transmitting useful light. The intent is to improve the contrast – the difference between the light and dark areas in the

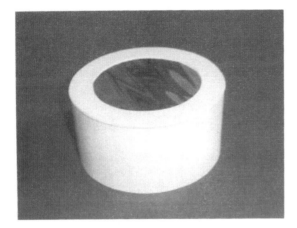

Figure 8.13. Baader AstroSolar Film in a homemade cell.

eyepiece. Most deep sky filters significantly darken the overall view but can make faint objects visible in moderate light pollution or under general skyglow. I do not recommend them in scopes smaller than 6in. There are four basic types of deep sky filters:

- **Broadband Filters** Allow most light to pass, but block wavelengths commonly produced by exterior lighting. They improve the contrast on most faint objects, but only slightly. Many people are very disappointed in these filters when using them visually, but they are effective in reducing skyglow in photographs. The LPR filter from Celestron, the Deep Sky filter from Lumicon, the SkyGlow filter from Orion, the LP-1 from Thousand Oaks, and the Moon and Skyglow filter from Baader Planetarium are broadband filters.

- **Narrowband Filters** Block much more light while passing the light emitted by many faint nebulae. They do not help with galaxies and star clusters. Examples are the Lumicon Ultra High Contrast, the LP-2 from Thousand Oaks, and the Orion UltraBlock.

- **Oxygen III (O-III) Filters** Block all but the one specific wavelength common to just a few nebulae – the Veil nebula and others. Examples are the Lumicon Oxygen III, the Meade Series 4000 Oxygen-III, and the LP-3 from Thousand Oaks.

- **Hydrogen Beta (H-Beta) Filters** Block all but the one specific wavelength common to just a few nebulae – among them the Horsehead and California nebula. The Lumicon Hydrogen Beta and the Thousand Oaks LP-4 are examples of H-Beta filters.

It should be noted that Lumicon is no longer in business, but their filters are still occasionally available at local astronomy stores and as used items on AstroMart or the classified ads on Astronomy-Mall (see Appendix A).

Minus Violet Filters

Minus violet filters are designed to tame the false color in achromatic refractors like the NexStar 60 and 80. False color is not too noticeable in the NexStar 60, except on the brightest objects. It is quite obvious in the short focal length NexStar 80. If you find the color on brighter objects objectionable, you might consider a minus violet filter from Baader Planetarium (available from Alpine Astronomical in North America). Baader calls their filter the Contrast-Booster – not only does it remove false color, but it also blocks several wavelengths of manmade light pollution and natural Moon-induced skyglow, resulting in increased contrast on many deep sky objects. When combined with their IR-Cut (infrared cut) filter, much cleaner images can be captured with a digital camera. In addition, I find the Contrast-Booster to be the perfect filter for Jupiter. Sirius Optics is also known for minus violet filters, however I find their filters cause slight distortion, particularly when imaging.

Finder Scopes

All NexStar telescopes except the NexStar 8/11 GPS are equipped with a red-dot unity finder that Celestron calls Star Pointer. A red-dot finder is well suited to pointing the scope at bright objects such as planets and alignment stars, but it does not allow you to see objects fainter than those visible with the naked eye. In fact, Celestron's Star Pointer projects the red dot onto a tinted plastic window that makes it difficult to see anything *but* the brightest objects. Since most will only use the finder during the initial alignment procedure, this is not usually a problem.

The Star Pointer is the most likely part of the telescope to eventually break or become lost. Similar red-dot finders are available from Orion and Apogee, or you might want to step up a level with Rigel's QuickFinder. The QuickFinder has a blinking feature and a bull's-eye rather than a dot, making it much easier to accurately center fainter objects. It does look a little odd though, as it stands 4 inches tall like a little tower on the top of your scope. Like the QuickFinder, the TelRad (Figure 8.14) projects a bull's-eye on its window. It is also quite large although it lies along the tube of the telescope more like a traditional finder. Due to its size it is really only practical on the 8- and 11-inch NexStars.

The NexStar 8/11 GPS comes with a traditional 9×50 straight-through finder. It is a good-quality finder that lacks just one needed feature: a quick-release mounting bracket.

Figure 8.14. NexStar 11 GPS with standard and TelRad finder.
Photo courtesy Dave Cole.

Celestron makes such a bracket, model number 51149-A, or an alternative is a similar bracket from Antares (model number F50DTB) available from many sources. During the alignment procedure, you might find it easier to use a red-dot or other unity finder. If so, I recommend leaving the standard finder in place and mounting the unity finder with two-sided foam tape as shown in Figure 8.14

Scope Covers And Cases

During storage, transport, and use, you will likely find it necessary to protect your scope from damage, moisture, and dust. If you store your telescope in the house, a bed sheet or large pillowcase may be all that is needed to keep dust off. If you store your telescope in a garage or other outside building, a little more protection will likely be called for. When storing your scope, do not cover it with plastic or other non-breathing material, as moisture will likely collect leading to corrosion and mildew. Cloth or paper covers are best, with elastic or simply string to hold the bottom closed. Bed sheets are still good material in this case, although some sewing may be required. Another alternative is a paper lawn bag available at hardware, gardening, and home improvement stores.

When traveling to a remote location with your scope, you might want a waterproof cover in case a quick rainstorm blows in. A great inexpensive cover is a barbecue cover – the size for 22in kettle barbecues is perfect. They are large enough to provide quick protection for all NexStars, even the NexStar 11. More than once mine has kept my scope dry while I disassembled it and loaded it back in the car. For a larger cover that also protects your tripod, look into the Desert Storm Cover, available from Anacortes Telescope.

For transporting, Celestron makes soft cases for the smaller NexStars and hard cases for the NexStar 5/5i/8/8i. Most Celestron retailers should stock them. Orion also sells soft cases for scopes of various sizes; it is best to speak with them directly to determine which case fits your scope. Cases and Covers manufactures excellent soft cases for all models of NexStar and will customize their cases as well.

For perhaps the ultimate in hard cases for the larger NexStars, consider the well-built models from Jim's Mobile, Inc. (JMI) and Pelican. JMI makes cases for the NexStar 4, 5/5i, 8/8i, and 8/11 GPS. Several of the models include wheels and the 8/11 GPS case even includes a quick-release bracket for the finder scope, allowing the case to be a little smaller by removing the finder during transport. Pelican cases are rugged and many trust them for airline shipment, but you must purchase foam separately from JMI. ScopeGuard Cases' Telescope Transport Cases are made for the NexStar 8/11 GPS and utilize the original packing foam. They provide great protection for shipping, but are the largest cases mentioned here.

If these cases are a bit too costly for your needs, consider a good-sized plastic tub with cover from the housewares section of your local store. Coupled with foam (good sources: Foam Order – http://www.foamorder.com and CustomFitFoam – http://www.customfitfoam.com), these variously sized storage containers provide good protection when taking your scope for a ride in your car or van. Another source for shipping cases is a military surplus store. If you can find a military transport case that fits the original shipping foam, you have yourself the makings of a bulletproof case (well, not literally).

For lightweight protection, some have turned to various sizes of wheeled luggage. The little NexStars will fit into a hard-shell golf transport case. Generous foam padding would be called for but you would likely be able to fit the scope still mounted on the tripod. Even with hard-side luggage and golf cases, I would not trust a delicate instrument like a scope to the airline baggage gorillas. Not to mention what airport security personnel are likely to do with a large, cylindrical metal object. When possible, hand-carry the optical tube (practical with the NexStar 80, 4 and 5/5i) and eyepieces and check the tripod in baggage.

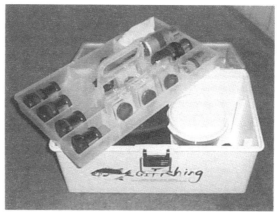

Figure 8.15. Tackle box accessory case.

Accessory and Eyepiece Cases

With many telescopes, you will likely only need a small case for eyepieces, filters, and a diagonal. With a computerized scope there are all those plus cables, batteries or power supply, a hand control and more. Sturdy plastic and aluminum eyepiece cases are available at many of the astronomy retailers listed in Appendix A. I would recommend steering away from the cases with cubed foam as the foam tends to shift around after you open a few slots for your growing collection. The cases with precut foam are better, although most foam-lined cases tend to shed tiny little pieces of foam that find their way onto your eyepieces and ultimately inside your scope.

With most of these astronomy-specific cases you need another carrier for your other accessories. Camera bags, the older style used when 35mm SLR cameras were more common, are great for larger accessories such as cables and the hand control. Plastic tubs and toolboxes are also great for larger parts. Many home improvement, hardware, and department stores carry large aluminum cases suitable for protecting astronomy gear. They are often sold as tool cases. Foam padding might be required if the case you select is unlined. Foam Order (http://www.foamorder.com) and CustomFitFoam (http://www.customfitfoam.com) are good sources for foam.

My personal favorite is a simple fishing tackle box (Figure 8.15). Most eyepieces fit in their own individual compartment in the trays designed for fishing lures. Other accessories can find homes in the bottom of the box or in unused compartments in the trays. Plastic tackle boxes do not scratch expensive accessories yet provide good protection. My entire complement of necessities fits in one tackle box.

Dew Prevention

Before long you will encounter a night with the right conditions to produce generous amounts of dew on your equipment. Dew forms quickly on any surface facing the sky. When dew accumulates on the optical surfaces of your scope, naturally it impedes your viewing.

Most refractors, including the NexStar 60 and 80, have small dew caps extending from the front of their objective lens. This dew shield reduces the lens's exposure to the sky, providing a good level of protection against the formation of dew. If you live in a dew-prone area, you might want to extend the length of the dew shield by wrapping the existing shield with foam. For extreme dew problems, consider one of the electrical dew prevention tools mentioned later.

The main optical surface of the NexStar 114 is the primary mirror at the rear of the tube. Thus, the optical tube is one very long dew shield. It is very rare for a Newtonian reflector to suffer from dew.

All other NexStar models are virtual dew magnets. The corrector plate on Maksutov and Schmidt–Cassegrain telescopes has no built-in protection from dew. One of the first items you must have for such a scope is a dew shield. Several different types of dew shields are manufactured for these scopes. Some are made of flexible plastic or foam to lay flat for storage while using Velcro® to hold their shape when used on the telescope. Others are rigid plastic or even aluminum cylinders. In practice, the flexible type is more convenient but some do not hold their shape well. Usually this can be remedied by attaching an appropriately sized hoop at the open end of the dew shield.

You can find dew shields at many of the astronomy retailers listed in Appendix A. Be certain the dew shield you purchase will really fit your scope. In particular, dew shields designed for older Celestron models might not fit the NexStar 8/11 GPS well, due to the side rails mounting the optical tube to the fork arms. In some cases, owners of these two models have used such dew shields by cutting material from them to allow them to fit around the mounting rails.

Dew shields have another benefit on Maks and SCTs: they block stray light from entering the optical tube from the side. Such light bounces around inside the tube and a small amount eventually reaches the eyepiece, reducing contrast or even producing glare. To maximize this benefit some dew shields are lined with flat black material to absorb light.

It is also fairly easy to make your own dew shield. For my NexStar 11 I constructed the 30-inch-long one shown in Figure 8.16 from a $\frac{3}{4}$ in camp pad. The dense foam is very sturdy and holds its shape well, and the insulating properties of foam also help prevent dew formation. To join the ends forming the cylinder, use a hot glue gun or self-adhesive Velcro®. Some try to make the cylinder tight enough for it to stay on the scope by itself, but I find the shield still tends to sag, so I use a luggage strap with a plastic snap buckle to secure the shield to the corrector cell.

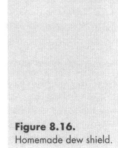

Figure 8.16.
Homemade dew shield.

Figure 8.17. Dew strip wrapped behind the corrector cell to allow secure mounting of dew shield. Photo courtesy Dan Hupp and Jeff Gerst.

Some nights dew will form even when a dew shield is used. One way to keep observing is to use a small hair dryer on a low setting to gently evaporate the dew. 12-volt warm-air guns are available from many astronomy retailers for use in the field. I don't really recommend this method; if it comes to the use of hot air to remove dew, you are fighting a losing battle: it will not take long for dew to form again, and you are baking dust and other contaminants onto your optics, making future cleanings all that more difficult.

A better method is to prevent the glass surface from cooling to the point of accumulating dew. This is done with an electrically powered heat strip. These dew strips wrap around the corrector (Mak/SCT) or lens (refractor) cell, are powered by a 12-volt power source, and provide enough heat to keep the glass above the dew point. You should always use a dew shield in concert with the dew strip for the best protection. With the NexStar 8/11 GPS, you may find it difficult to wrap the dew strip around the corrector cell and then mount the dew shield. In that case, wrap the dew strip around the carbon fiber tube, directly behind the corrector cell as shown in Figure 8.17. This should allow the dew shield to fit normally.

The two most commonly used dew strips are from Orion and Kendrick Astro Instruments. The Orion models are much cheaper, but they run continuously and quickly drain a 12-volt battery. The Kendrick models require a controller unit that allows variable power settings. The highest setting keeps the dew strip running continually while the lower settings cycle the power on and off at various timing cycles. This greatly improves battery life, but it is pretty much a guessing game to determine the optimal setting for an evening. In addition to dew strips for the corrector or main lens, the Kendrick controller can also drive additional optional dew strips for eyepieces, finder scopes, and more.

A newer controller was recently introduced. The DewBuster by Ron Keating (http://www.gbronline.com/ronkeating) cycles the power on and off according to temperature. This takes the guesswork out of the optimal power setting. It was designed as a direct replacement for the Kendrick controller and in fact you can use the Kendrick dew strips with DewBuster. The price is higher, but worth it in areas with heavy dew formation. Ron's web site even includes plans for homemade dew strips.

Cold Weather Gear for Your Scope

Cold weather is not friendly to an electronically controlled telescope. If you resisted buying a large rechargeable power source, you will discover that your little AA batteries run out of juice even faster in cold weather. An a.c. adapter or other good power source is a necessity in winter weather.

Additionally, the LCD panel on the hand control becomes sluggish and even illegible in cold weather. If it isn't too cold, you can slow down the rate for the scrolling information using the 1, 4, and 7 buttons as described in Chapter 5. But when the temperature really drops, your hand control needs some heat to operate normally. If you are already using the Kendrick or DewBuster dew prevention systems, Kendrick makes a heater element for hand controls. The Kendrick TelRad heater element also works well for hand controls. Another option is chemical hand warmers available from most outdoor equipment suppliers. Attach one to the back of the hand control with elastic.

If your hand control is past warranty and you are handy with a soldering iron, another solution is to mount two small resistors inside the hand control. Power for these mini heaters comes from the hand control itself. Visit the Astro Articles page at Matthias Bopp's web site (http://www.dd1us.de) and read the article for the N5/8 or the N5i/8i electronic hand control heater.

While dew can be a problem at other times of the year, frost will form on optics in the winter. Powered dew prevention strips discussed earlier are your best defense against frost.

Mounting Rails and Counterweights

At some point, you might like to mount extra equipment such as cameras on top of your scope or you will need counterweights to balance heavy equipment mounted at the rear of the scope. The NexStar 5/5i/8 should be balanced a little front-heavy for best tracking, while the NexStar 8i and NexStar 8/11 GPS track best when back heavy. Mounting rails with sliding counterweights provide great flexibility when balancing a fork-mounted scope, but only the NexStar 8/11 GPS are drilled for rails. Another popular use of rails on the NexStar 8/11 GPS is to mount small refractors on top to provide a wide-field view.

Losmandy makes premium rails and mounting accessories for a wide variety of uses. Less expensive, but sufficient for all but the heaviest of equipment, are the mounting rails from ScopeStuff. In addition to equipment, their web site has an incredible amount of

Figure 8.18.
Homemade mounting rail. Photo courtesy Robert Cadloff.

information at http://www.scopestuff.com/ss_mrail.htm. You should also visit Robert Cadloff's web site – http://www3.sympatico.ca/bomo/astro/jig/railsystem.html. Robert provides simple instructions for making your own mounting rails (Figure 8.18) and counterweight systems, including links to good sources for raw materials.

In addition to Robert Cadloff's web site and ScopeStuff, visit Starizona's web site for counterweight sets and a wealth of information about balancing a fork-mounted scope – http://www.starizona.com/basics/balance.html. Their rails also are suitable for mounting equipment.

One of the best options for balancing the NexStar 5/5i/8/8i is Ray's Bracket sold at BuyAstroStuff.com. This bracket allows the optical tube to be slid fore and aft to improve balance. Ray's Bracket will be discussed fully in Chapter 13.

If your needs are simply to mount a camera piggyback, Celestron makes the model 93598 piggyback camera mount for the NexStar 8/8i and 8/11 GPS and model 93601 for the NexStar 5/5i. ScopeTronix sells piggyback camera mounts for the NexStar 4, 5, and 5i.

Observing Chairs

Other than warm clothes in winter, no other item will afford as much observing comfort as a well-designed chair. The ultimate chair for use at the telescope should adjust height with ease, fit easily between the tripod legs, sport comfortable padding, and fold into a small profile for ease in transport and storage. The most common design to meet these requirements has an inclined rail that the supports the seat at a wide variety of heights. Many astronomy retailers carry such a chair, but one stands out – the Astro Chair from BuyAstroStuff.com. It is very well built and costs much less than competing chairs. With only one hand, the padded seat adjusts in height from 18 to 31 inches, sufficient range for use with all NexStar models (Figure 8.19).

If your woodworking skills are better than mine, you can even build your own chair. There are several common designs; one such is the Denver Observer's Seat – http://members.tripod.com/denverastro/seat.html.

Other alternatives are pneumatic chairs similar to office chairs without the back. Shutan and others sell such a model. In fact, if your observing location has a concrete or solid wood surface, an office chair might be the best alternative for you. Rolling around on the wheels would be a wonderful convenience if you can find a chair that adjusts high enough.

Field Tables

A lot of people don't seem to consider this accessory important, but I couldn't live without mine. The first thing I set up when observing is my table. My accessory case then sits on top for the rest of the night. I then have a place for clipboards, books, pens, and more important things like drinks and snacks. Any table will do, but portability should be considered. The outdoor equipment suppliers listed in Appendix A stock a variety of tables designed for camping. These include roll-up tables and several types of folding tables. My personal favorite is the Coleman "Tailgater" folding table. At 24 by 48 inches, it is large enough to hold plenty yet it folds into a 24-by-24-by-3-inch package that fits easily in most cars. You might already have something suitable – card tables and TV trays make good traveling scope-side tables.

The NexStar User's Guide

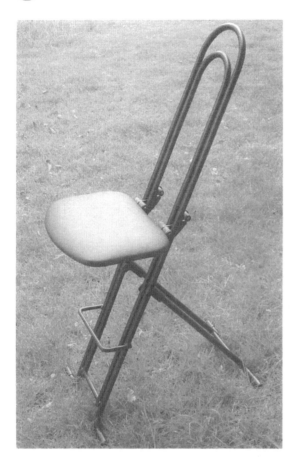

Figure 8.19.
BuyAstroStuff.com's
Astro Chair.

Hand Control Extension Cables

If you find the coiled cord on the hand control is just too short for you, it is fairly easy to extend the length. You need a 6-conductor cable with an RJ-12 plug (6 gold leads) on each end and an RJ-12 or RJ-45 coupler that will allow you to connect the extension cable and the hand control cable together. Naturally it is important that the wire path remain "straight through", meaning pin 1 on one end goes to pin 1 on the other end, pin 2 to pin 2, etc. If you cross the connections it is likely that you will damage your hand control! It is easy enough to make one of these if you have the right materials and tools, but the following are available from Radio Shack (http://www.radioshack.com) for less than $15:

- 25ft Line Cord (Flat), part #: 279-422
- 6-Conductor In-Line Coupler, part #: 279-423

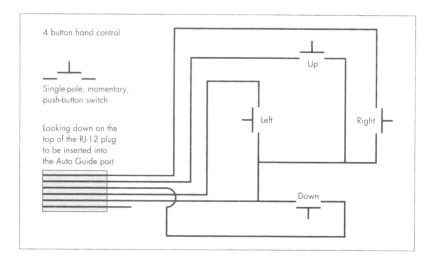

Figure 8.20. Foour-button hand control schematic.

Manual Hand Control for the N8/11GPS and N5i/8i

A manual hand control can be attached to the Auto Guide port on the NexStar 8/11 GPS and NexStar 5i/8i. This is very useful for manually guiding the scope during long-exposure astrophotography. When using the normal hand control to guide, you may accidentally press the Enter button, causing the telescope to swing away and re-center, ruining your exposure.

If you can find it, you can purchase the manual hand control (Celestron part number 28983) made for the CI-700 and some of the older Celestron scopes. You can also build one yourself. The required parts are a small project box, four normally OFF push-button (momentary) switches, and an RJ-12 cable. Refer to Figure 8.20 for the correct wiring.

For further reference, the pins in the Auto Guide port on the base of the telescope are numbered 1 through 6 going left to right as shown in Figure 8.21. The function of each pin is also shown in Table 8.5. If pin 3 is electrically connected to pin 2, the telescope moves left, pin 4 to pin 2 moves down, etc.

Table 8.5. Functions of pins in Auto Guide port

Auto Guide port pin	Function
1	Not connected
2	Ground
3	Move left/east
4	Move down/negative declination
5	Move up/positive declination
6	Move right/west

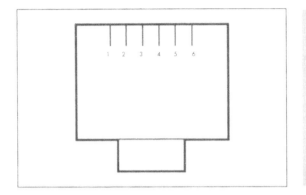

Figure 8.21. Auto Guide port on the NexStar 5i/8i and NexStar 8/11 GPS.

If tiny switches, soldering irons, and wire don't excite you, Mike Zeidler sells a complete 4-button hand control (Figure 8.22) for a very reasonable price. You can contact him via email at zeidler@igs.net.

Conclusion

Actually, there is no conclusion to this chapter ... you will likely find that once you are infected with "equipmentitis" it will be hard to stop looking for new astronomy equip-

Figure 8.22. Manual hand control. Photo courtesy Mike Zeidler.

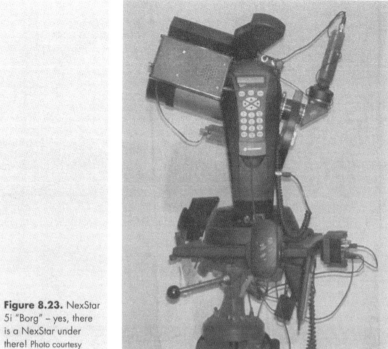

Figure 8.23. NexStar 5i "Borg" – yes, there is a NexStar under there! Photo courtesy Matthias Bopp.

ment. But you will never be justly accused of being obsessed until your telescope begins to resemble Matthias Bopp's NexStar 5i "Borg" (Figure 8.23)! If you are curious about all the gear burdening that poor little NexStar, visit Matthias' web site – http://www.dd1us.de.

In this chapter I suggested several useful accessories for your NexStar telescope, but there are more out there. If I didn't discuss enough ways to help you spend your money, I would refer you to Phil Harrington's book *Star Ware*, a comprehensive look at astronomy equipment for the amateur astronomer.

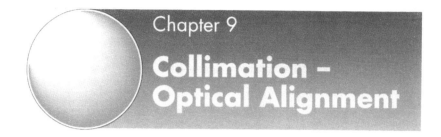

Chapter 9

Collimation – Optical Alignment

Collimation is the process of aligning all the optical components of a telescope so that the focal plane of the telescope meets the focal plane of the eyepiece squarely. With any telescope there are many potential sources for misalignment, but only a limited number of components are easily adjusted. In fact, some telescopes do not have any user adjustments at all. These scopes are collimated at the factory or the design of the telescope itself insures collimation.

Collimation is critical for good optical performance. A poorly collimated telescope does not show as much detail as when it is well collimated. For example, on Jupiter, accurate collimation can be the difference that enables us to see the Great Red Spot and shadow transits of the Jovian moons. SCT and Newtonian telescopes require very accurate collimation and they often fall out of collimation after traveling or after a rough jarring. For these scopes, you should get in the habit of checking collimation regularly. For refractor and Maksutov telescopes, testing the scope once after you receive it should be sufficient, as they hold collimation well.

The general test for collimation requires setting up your scope and waiting for it to cool to the ambient temperature. This might take as long as an hour for the NexStar 4 and the SCT models, but for the other NexStars 20 minutes should be sufficient. Align the scope so that it is tracking the sky and GoTo a 2nd-magnitude star – Polaris is a good choice.

Select an eyepiece that gives a magnification about two times the aperture size in millimeters – for example, 120× for the NexStar 60 or 560 × for the NexStar 11 GPS. *Slightly* defocus the image just enough to see a set of three or more rings. If you see a set of perfectly concentric circles, your scope is fairly well collimated. If the circles are offset to one side, adjustments are in order. Refer to Figure 9.1 for comparison, but note that on Newtonian, Maksutov, and Schmidt–Cassegrain telescopes the center of the image will be darkened by the shadow of the secondary mirror. If the circles are concentric, you won't need to worry about this chapter just yet, although with the procedures described here it may be possible to improve the collimation if seeing conditions are very good.

It is very important to note that poor seeing conditions do not allow precise collimation. In fact, it is likely that you will worsen the state of your telescope if you attempt collimation with a shaky, wavering image. The same holds true with a telescope that has not yet cooled sufficiently. Currents of air inside the optical tube will deform the out-of-focus circles in a way that makes the image look like the scope is out of collimation. As you attempt to align the optics, you never seem to make the correct adjustments and ultimately

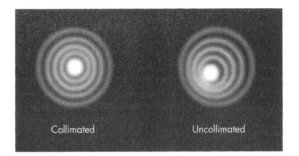

Figure 9.1.
Unfocused star images through a collimated and an uncollimated telescope.

the situation just gets worse. If you cannot see stable, clearly visible rings, wait for another night to collimate.

If it seems you never get an evening with good seeing conditions, you might want to try collimating in the day with an "artificial" star. Look for a shining reflection of sunlight off in the distance, perhaps a steel fixture on a distant building or a glass insulator on an electric pole projecting a glint of sunlight. Or you can construct an artificial star with a ball bearing or a glass Christmas tree ornament. Place it where sunlight glints from its surface and far enough away for the telescope to be able to focus on it, preferably with a field of grass between – asphalt and concrete radiate heat in the sunlight and produce unsteady seeing conditions. You will find that the out-of-focus image looks surprising like that of a star.

Another potential aid, particularly for Newtonian scopes like the NexStar 114, is a laser collimator. In particular, holographic laser collimators, with their projected pattern, make accurate collimation of a Newtonian telescope much easier. The holographic lasers from Helix Manufacturing (http://www.helix-mfg.com) and Howie Glatter (http://www.collimator.com) are two fine examples. To aid in precise final collimation using a defocused star, you might consider the Star Collimator from EZTelescope (http://www.eztelescope.com). This unique eyepiece has concentric reticle rings that allow easy centering of the defocused star pattern. This typically allows a much more precise collimation at a lower power and thus you are not as limited by poor seeing conditions.

The actual procedure for collimation varies according to the telescope design, so we will discuss each separately.

SCT Collimation

(NexStar 5 and Larger)

The only collimation adjustment for most Schmidt–Cassegrain telescopes (SCT) is the secondary mirror at the front of the telescope. Due to the magnifying nature of the secondary mirror in an SCT these scopes are very sensitive to small inaccuracies in collimation. Fortunately, commercially produced SCTs are also the easiest type of scope to collimate.

Collimation is accomplished by adjusting the three screws on the secondary holder at the front of the telescope. As you can see from Figure 9.2, the screws cause the secondary mirror to tilt on a rocker pivot. Adjusting the screws requires either a cross-tip screwdriver, preferably of short length, or a hex wrench. Be certain it is the correct size to seat properly in the screw heads. Tightening one screw will generally require loosening the other two screws by no more than half the same amount.

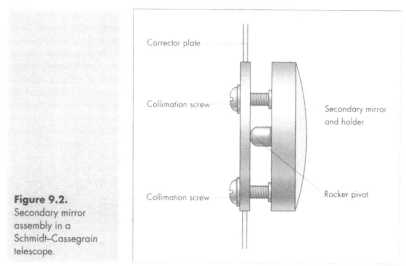

Figure 9.2.
Secondary mirror
assembly in a
Schmidt–Cassegrain
telescope.

Having a pointed metal object so close to the delicate corrector plate in the dark is a little nerve-racking for most. One of best investments you can make for a Schmidt–Cassegrain telescope is a set of no-tools, thumbscrew collimation screws. The added convenience makes collimation a quick, simple procedure. The most popular are Bob's Knobs from Morrow Technical Services or the collimation screws available from ScopeStuff. When installing such screws note that on some Schmidt–Cassegrain telescopes the only things holding the secondary mirror in place are the collimation screws. Point the front of the telescope downwards, loosen all three screws about a half turn, then remove and replace them one at a time. Removing all three at the same time could release the secondary mirror from its perch with catastrophic results.

To start collimation, align the scope to start tracking. GoTo or manually slew to a 2nd-magnitude star – 3rd-magnitude is often best for an 11-inch scope – above about 70° altitude. If your telescope is severely mis-collimated, you may need to start with a brighter star. Select an eyepiece that gives a magnification of about the same as the aperture in millimeters, for example 200× for the NexStar 8. Accurately center the star in the field of view. Defocus slightly until concentric circles with a small, dark shadow in the middle are clearly visible.

Viewing this donut-shape, place your finger in front of the corrector plate and move it around the edge until you see the shadow of your finger (or hand if you can't see the shadow of your finger) at the thin part of the donut. Be careful not to touch the corrector plate. Look at the front of the scope and tighten the collimation screw closest to your finger. If the screw is very tight, loosen the other screws a bit. If your finger lies between two screws, tighten them both a bit. Look in the eyepiece and re-center the star. It is **very** important to re-center the star after each adjustment; creating perfectly concentric circles on a star not centered in the eyepiece results in a telescope that is **out** of collimation. Repeat this process until the concentric circles are centered. Final adjustments might require just a little pressure on the screws, with hardly any movement noticed.

If the seeing is good, increase the magnification to two times the aperture in millimeters and defocus just enough to see three or four tiny concentric rings. Follow the previous procedure to center these rings. If the seeing is exceptional, focus the star and look for the Airy disk and the one or two diffraction rings around it. If the scope is perfectly collimated, the diffraction rings will run all the way around the Airy disk. If not, the diffraction ring will be broken. Adjust for a complete diffraction ring around the Airy disk and things are as good as they get.

If the final collimated position results in screws that are slightly loose, the telescope will not hold collimation. It is best if your final adjustments are tightening only, but don't force anything. This is another advantage of the no-tools thumbscrews; your fingers provide much better feedback than a screwdriver. Additionally, with the NexStar 5/5i you can adjust these thumbscrews while observing the results in the eyepiece.

Despite how complicated this might sound, it really is simple. After a few goes of it, you will find you can do it without even thinking much about the process. Remember though, if the telescope is not adequately cooled or the seeing conditions are not stable, you will not see steady rings and you should not adjust collimation. Wait for a better night.

Newtonian Collimation

(NexStar 114)

Of all the common types of telescopes, a Newtonian is the most complicated to collimate. This is mostly due to the relatively complex optical alignment of the secondary mirror and the several different ways it can become misaligned. The critical factors for the secondary mirror are:

1. It must be accurately centered in the main tube.
2. It must be directly under the focuser tube (or offset a little towards the primary mirror end of the main tube).
3. The flat of the mirror must be aligned with the focuser tube axis, not tilted above or below the view in the focuser tube.
4. It must be angled so that the focuser tube axis (the optical center of the inserted eyepiece) reflects off the secondary and strikes the exact center of the primary mirror.

To further complicate matters, the focuser tube must be square with the main tube.

For this discussion, we will assume that the factory did their part by centering the secondary in the main tube and constructing a focuser tube that is square with the main tube. If you suspect this is not the case, you might use thin cardboard and a razor knife to construct templates to check and adjust as necessary.

One additional complicating factor is the corrector lens at the bottom of the focuser tube. This makes it difficult to get the rough mechanical alignments necessary prior to aligning with a star. To check that the secondary is directly under the focuser tube and aligned with the focuser tube axis, we must remove the corrector lens. If your scope is very close to accurate collimation, as indicated by the star test at the beginning of this chapter, these steps may not be necessary and you can skip to the final adjustments section below.

Before You Start

Prior to beginning, you either need to purchase a sight tube such as Orion's Collimating Eyepiece or you can make a similar tool using a 35mm film canister. Using a solid-colored film canister, not clear plastic, cut and remove the bottom. Make a very small (1mm or smaller) hole in the precise center of the canister's cap. If it is not precisely centered or if it is too large, it will not work correctly. One good method for creating the hole is to use a large safety pin spread open. First, use the tip of the pin to make a pin prick at the exact center. Then, use a pair of pliers to hold the pin and heat it in a flame until red-hot. Then you can easily push it through the center of the cap.

You will also need:

- a cross-tip screwdriver;
- a hex wrench for the screws on the secondary mirror;
- a piece of paper;
- a pair of scissors;
- a dark, fine-tip marker – a permanent marker or a dry-erase marker are fine, but a water-based marker typically found in an arts supply store will not do;
- recommended if using a laser collimator – a self-adhesive paper hole reinforcement ring ("paper donut").

First, you must remove the corrector lens. Rack the focuser all the way out and loosen the two screws securing the focuser knob shaft enough to allow the chrome focuser tube to slide completely out. At the bottom of the focuser tube you will find a plastic retainer ring with two notches. Using your thumb and fingernails or soft plastic implements, unscrew this ring. **Important:** Note which side of the lens faces the bottom of the tube. If you re-install the lens backwards, the scope will **not** focus. It is best to make a note. Set the lens aside on a soft, clean cloth. Replace the focuser tube and tighten the screws for the focuser knob shaft.

Next, you must remove the primary mirror and mark the precise center of the mirror. Remove the three screws on the sides of the primary mirror cell, the black metal ring on the rear of the scope. Be careful to hold the mirror cell as you remove the screws – we don't want the primary mirror to go crashing to the floor! Do not touch the surface of the mirror; it is very delicate and easy to scratch. Remove the screws on the clips holding the mirror in the cell. Set the mirror, shiny side down, on top of a piece of paper and trace its outline onto the paper. Set the mirror, shiny side up, to the side.

Cut the circle from the paper. Fold the circle in half, then in half again to produce a wedge. Cut about $\frac{1}{8}$ in (3 mm) from the tip of the wedge to produce a small hole in the exact center of the circle. Center the opened paper over the mirror and gently mark the center of the mirror with a good-sized dot. If you like, you may center a self-adhesive hole reinforcement ring around the dot to make it easier to see. Do not be concerned about these marks in the center of the primary mirror; no light from the sky reaches this part of the mirror as it is in the shadow of the secondary mirror.

Replace the primary mirror in the mirror cell, tightening the clips enough to prevent the primary from rotating or moving, but not too tight or you will distort the shape of the mirror, resulting in "pinched optics". Pinched optics are recognizable when you are viewing the out-of-focus star image. Rather than round concentric circles, you will see circles with three corners. Set the mirror cell aside for now.

Coarse Alignment of Secondary and Primary

The various reflections of the two mirrors presents a view that most initially find confusing. Figure 9.3 should help you identify the various parts you see through the peephole. If you are not using a sight tube with crosshairs, then the two crossed lines will be missing from your view of the sight tube and you will likely see only the center of your eye in the **reflection** of the secondary (the last object in Figure 9.3). If you find yourself loosing track of what you are trying to center, refer back to this figure.

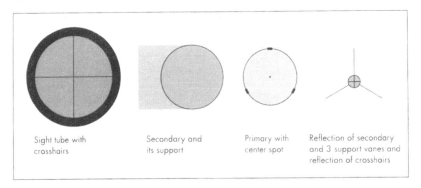

| Sight tube with crosshairs | Secondary and its support | Primary with center spot | Reflection of secondary and 3 support vanes and reflection of crosshairs |

Figure 9.3. Various objects viewed through the focuser tube of a Newtonian telescope.

To start, insert the sight-tube or film canister squarely in the focuser tube. Point the back of the main tube at a well-lit, light-colored wall. Look into the peephole and you will see the secondary mirror. Look at the mirror itself, not the reflection of the mirror. The circle of the mirror face itself should be centered in the peephole. You will find it easier to gauge if you rack the focuser tube in or out to fill the peephole with the secondary mirror. If the secondary is not centered, loosen the center screw on the secondary holder and adjust the mirror until it is centered and facing directly towards you. Refer to Figure 9.4 to confirm the correct view. Retighten the center screw.

Next we will adjust the tilt of the secondary mirror to point directly to the center of the primary mirror. Adjust the three hex screws on the secondary mount to roughly center the reflection of the open tube on the face of the secondary mirror. This is usually easiest if you only adjust two of the screws, leaving the third alone unless one of the other two reaches the end of its adjustment. Replace the primary mirror cell and secure it with the previously removed screws. In the peephole you will see several reflections, but you need to concentrate on the reflection of the primary mirror. If you are using a sight tube with crosshairs, adjust the three hex screws on the secondary to center the cross directly over the dot in the center of the primary mirror. If you are using the film canister, you will adjust the hex

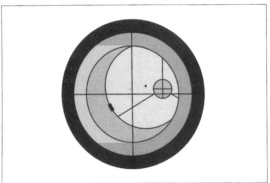

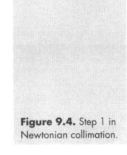

Figure 9.4. Step 1 in Newtonian collimation.

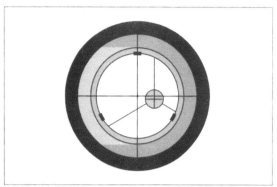

Figure 9.5. Step 2 in Newtonian collimation.

screws to center the reflection of the entire primary mirror in the secondary itself. Refer to Figure 9.5 for an example of the correct alignment.

The hardest part is now behind you. The secondary mirror should remain well aligned unless the scope is severely jarred or the screws are adjusted.

The last step in this mechanical alignment process is to adjust the tilt of the primary mirror so that the light reflected from it comes right back to the center of the focuser tube. The rear of the primary mirror cell has three cross-head screws and three thumbscrews. The cross-head screws lock the primary mirror in position and the thumbscrews adjust the tilt. Loosen the cross-head screws about two turns. Then, looking in the peephole, adjust the thumbscrews – again, adjusting just two is best – until the reflection of the focuser tube and the peephole (you might be able to see your eye) are precisely aligned with the center dot on the primary. If a thumbscrew will move no further, loosen the corresponding cross-head screw. Refer to Figure 9.6 for this confusing view. Tighten the cross-head screws to fix the mirror in place. Note that this will sometimes shift the primary mirror slightly and some further adjustment with the thumbscrews may be necessary.

With this accomplished, loosen the two screws holding the focuser knob shaft and remove the chrome focuser tube. Replace the corrector lens (be sure you get it right side up – remember, if you can't even focus, this is most likely the culprit) and tighten the retaining ring with your fingernails or soft plastic implements. Replace the focuser shaft and tighten the two screws enough to produce smooth focusing motion. Too tight and the

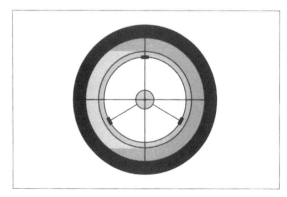

Figure 9.6. Step 3 in Newtonian collimation.

focuser will require too much effort to move smoothly; too loose and the focuser will easily slide in when you press the end of the tube.

Final Adjustments

The finishing adjustments necessary to produce the best images from your telescope require collimation with a star. All final adjustments should be to the **primary mirror only**, although an exception will be discussed at the end of this section. To start collimation, align the scope to start tracking. Wait at least 20 minutes for your scope to cool to the ambient temperature. GoTo or manually slew to a 2nd-magnitude star above about 70° altitude. Select an eyepiece that gives a magnification of about 100× – a 10mm eyepiece. Accurately center the star in the field of view. Defocus slightly until concentric circles with a small, dark shadow in the middle are clearly visible.

If the rings are concentric, skip to the next paragraph to see if you can improve the collimation as shown at a higher magnification. If the rings are not concentric, loosen the three cross-tip locking screws on the back of the primary mirror cell by half a turn. Then, while observing the rings in the eyepiece, very slightly turn one of the thumbscrews on the primary mirror cell. This will either improve or worsen the concentricity of the rings. With careful adjustments to just two of the thumbscrews you should be able to center the rings. If a thumbscrew will move no more, loosen the corresponding cross-head screw slightly. Look in the eyepiece and re-center the star. It is **very** important to re-center the star after each adjustment; creating perfectly concentric circles on a star not centered in the eyepiece results in a telescope that is **out** of collimation. Repeat this process until the concentric circles are centered. Final adjustments might require just a little pressure on the screws, with hardly any movement noticed.

If the seeing is good, increase the magnification to 200× – a 10mm eyepiece with a Barlow lens is good. Follow the previous procedure to center these rings. If the seeing is exceptional, focus the star and look for the Airy disk and the one or two diffraction rings around it. If the scope is perfectly collimated, the diffraction rings will run all the way around the Airy disk. If not, the diffraction ring will be broken. Adjust for a complete diffraction ring around the Airy disk and things are as good as they get.

After you are satisfied with the results, tighten the three cross-head screws to lock the primary mirror in place. Note that this might somewhat alter collimation and it would be necessary to slightly adjust the thumbscrews again.

Remember, if the telescope is not adequately cooled or the seeing conditions are not stable, you will not see steady rings and you should not adjust collimation. Wait for a better night.

There is one final test that you should make. Look at the concentric rings as you rack the focus knobs in and out past the actual focus point. The rings should be concentric both inside and outside of focus. If they are not, it either indicates less than perfect optics or your centering and aiming of the secondary mirror in the previous section was not as accurate as it might have been (if, for example, your pinhole in the film canister was not accurately centered). It is also possible that the secondary mirror is not centered in the tube itself or the focuser is not square to the main tube (refer to the beginning of this section on Newtonian collimation for the complicating factors in Newtonian scopes). Some have had success reaching a more accurate collimation by looking at the side of focus with the non-concentric rings and adjusting the hex screws on the secondary to make the rings concentric. Then it is necessary to defocus the other direction and adjust the primary mirror to center the rings. Keep in mind that you **must** re-center the star after each adjustment.

Understand that it is possible to make things even worse this way and then you will be forced to start over with the coarse adjustments.

My last bit of advice on collimating a Newtonian scope is to find an experienced amateur astronomer to assist you if you run into problems. The first few times you collimate a Newtonian scope can be quite trying.

Maksutov Collimation

(NexStar 4)

The NexStar 4 is not designed for user collimation. It is adjusted at the factory and Maksutov telescopes do a good job of holding collimation. However, some NexStar 4s have been found to be slightly out of collimation. If you test your NexStar 4 as described at the beginning of this chapter and find it needs collimation you can either send it to Celestron or you can collimate it yourself. Be warned, though: since it was not designed for this, it can be a little frustrating.

Start by removing the rubber focus knob, loosen the setscrew found underneath, and then remove the metal bushing. Loosen and remove three of the screws holding the plastic back onto the optical tube. Align the scope to start tracking. Wait at least one hour for your scope to cool to the ambient temperature. GoTo or manually slew to a 2nd-magnitude star above about 70° altitude. Select an eyepiece that gives a magnification of about 100× – a 13mm eyepiece or the closest you have. Accurately center the star in the field of view. Defocus slightly until concentric circles with a small, dark shadow in the middle are clearly visible.

If the rings are concentric, skip to the next paragraph to see if you can improve the collimation as shown at a higher magnification. If the rings are not concentric, hold your finger in front of the corrector lens (be careful not to touch it) and move it around until it the shadow of your finger is located at the thinnest side of the rings. Note this location. Loosen and remove the remaining screw and remove the plastic back with the eyepiece. Notice how the mirror lever engages the flip mirror – you will need to put it back together in a minute! Be very careful not to touch the mirror or you may scratch it. After removing the back, you will find three pairs of screws – each pair consisting of a cross-tip and hex screw. The cross-head screws pull the primary mirror, while the hex screws push. Recalling the location where your finger found the thinnest part of the rings, loosen the cross-head screw on that side of the scope **very** slightly and tighten the hex screw. Replace the back of the scope (watch that mirror lever), re-center the star in the eyepiece, and check the defocused rings. Continue adjusting until you see concentric rings. Final adjustments might require just a little pressure on the screws, with hardly any movement noticed.

If the seeing is good, increase the magnification to about 200× – a 13mm eyepiece with Barlow lens is good. Follow the previous procedure to center these rings. If the seeing is exceptional, focus the star and look for the Airy disk and the one or two diffraction rings around it. If the scope is perfectly collimated, the diffraction rings will run all the way around the Airy disk. If not, the diffraction ring will be broken. Adjust for a complete diffraction ring around the Airy disk and things are as good as it gets.

Reassemble your telescope and enjoy the improved views. Your NexStar 4 should hold collimation indefinitely.

Refractor Collimation

(NexStar 60 and 80)

Most refractors, the NexStar 60 and 80 included, are not designed for adjustable collimation. The plastic cell that holds the objective lens is manufactured to hold the lens assembly at a perpendicular orientation to the main tube and it has no adjustments. But it is possible that, when performing the star test outlined at the beginning of this chapter, you may discover your telescope is out of collimation.

A loose retaining ring at the front of the lens cell almost always causes this condition. The fix is quite easy. Point the front of the scope upward and remove the dust cover and dew shield. With a clean, lint-free cloth or tissue, press down all around the outside edge of the lens. Then tighten (turn clockwise) the inner retaining ring just until it is snug.

This ring can also be too tight. When this happens, the defocused star image will display slight corners – the sign of pinched optics. This can occur during cold weather as the contracting plastic presses down on the lens. Slightly loosening (turning counterclockwise) the retaining ring will solve the problem.

Chapter 10
Controlling Your NexStar with a PC or Palmtop Computer

Although the computerized hand control performs all essential functions that are needed for a night out under the stars, many NexStar owners also elect to connect a personal computer (PC) or palmtop computer to their telescope for added features and convenience. And then there is just something about computers and the crowd that purchase GoTo scopes ...

A planetarium program, discussed in Chapter 2, can show you the section of sky currently in the eyepiece of the telescope, helping you to identify objects you are currently observing. You can select an object in that same planetarium program and your PC can direct the telescope to GoTo that object. This makes it very easy to work through several objects in the same area of the sky. Other programs, such as NexStar Observer List, discussed in Chapter 7, can help you plan an observing session. Then when you are out under the stars, the program can direct your NexStar to GoTo each object on your list, one after another.

A PC or palmtop program can easily allow GoTo for objects not in the hand control's database, for example, asteroids and comets. And at least one program, Satellite Tracker, will allow you to view satellites and the International Space Station with NexStar telescopes. Smooth tracking of satellites is possible with the NexStar 8/11 GPS and 5i/8i. Other models are capable of "leapfrog" tracking – the scope jumps ahead and waits for the satellite to pass through the field of view. Then Satellite Tracker leaps forward for the next pass through the eyepiece.

All models of NexStar telescopes are capable of interfacing with a PC or palmtop computer. With the NexStar 5i/8i, the optional computerized hand control is required. All other NexStar models are ready out of the box. In all cases you will need a connection cable, a suitable PC or palmtop computer, and software to run on that computer. The cable connects to the bottom of the NexStar hand control – the RS-232 jack.

There are some details you should be aware of with the little NexStars (the 60/80/114/4 models). The NexStar 80 and 114 were sold in an "HC" configuration with a manual hand control and a "GT" configuration with a computerized hand control. The HC hand control is capable of PC control, but only with the supplied GuideStar software from Celestron. This software allows a PC to emulate the computerized GT hand control, thus providing complete GoTo capability. Also, a supplemental software package, Arrow Keys for

GuideStar, has been developed by Michael Ganslmeier and Matthias Bopp to add on-screen arrow keys to the GuideStar program. Arrow Keys For GuideStar is available for free download from http://www.dd1us.de.

Also, as mentioned in Chapter 3, there have been two versions of the GT hand control for the NexStar 60/80/114/4 telescopes. These two versions use different control commands. Thus, a program that is compatible with the new GT hand control may not be compatible with the original GT hand control. Also, the original GT hand control has several bugs in the PC control routines, of which two are most notable. First, after the scope finishes a slew after sending a GoTo command from your PC, you **must** move the scope manually with the arrow keys on the hand control or the hand control will become unresponsive to future GoTo commands. Second, when the PC retrieves the RA–Dec coordinates from the scope, the original GT hand control reports incorrect coordinates for more than half of the sky. Refer to Chapter 3 if you are not sure which version you have.

What Types of Computers Work

Potentially, any computer with the ability to use a serial port – also known as an RS-232 port – is capable of controlling a NexStar telescope. This includes almost all desktop or laptop (notebook) computers running MS-DOS, Microsoft Windows, the Macintosh OS, or the various flavors of Unix. This also includes most palmtop computers running the Palm OS or Windows CE.

When searching for a laptop or desktop computer, you will find it more convenient to use a model with a built-in serial port. Many newer laptop computers do not have serial ports. In that case, you must purchase a USB-to-serial adapter (about $40) to provide the required serial port. Other nice features include:

- a large keyboard for use with gloved hands;
- extra long battery life or a 12V cigarette-lighter power adapter option;
- a fast processor, lots of memory, and lots of hard drive space – especially if you intend to try your hand at digital imaging.

With palmtop computers, be sure to get a model that offers a **serial** sync cable (a USB sync cable will not do) or you will not be able to connect it to your telescope without an expensive compact flash serial adapter card. With most palmtop computers, the serial cable is optional; generally only a USB cable is included. Other nice features are:

- a color display – much easier to read in the dark;
- a backlit display or a clip-on light (don't forget to color the bulb red!);
- long battery life or a 12V cigarette-lighter power adapter option;
- extra memory for larger-object databases in your astronomy software.

Although most astronomy software includes a "night-vision" mode that changes the screen to shades of red, on most computers this will still be too bright. Try using one or two sheets of dark-red plastic instead. Check plastic suppliers, art stores, and camera shops for plastic sheets. It will need to be darker than you might think; try stacking sheets to get the right density. A simple cardboard frame secured to the computer with Velcro will keep it all in place.

You will also find it hard to type on an unlit keyboard. A small, red LED light attached to the top of the display will fix that. Clip-on lights, typically marketed for use in airplanes, will generally be too bright without modification. The key is to use a faint, red light to preserve your eyes' dark adaptation.

Power in the field will generally be a problem. The batteries in most laptops and palmtops do not last long, particularly when using the serial port and potentially capturing images. If you are not near an a.c. power outlet you will most likely be using a 12V battery with a cigarette-lighter socket. Some laptop and palmtop computers have a 12V power adapter option. For others you will need a d.c.-to-a.c. power inverter. Widely available in automotive parts stores, a power inverter provides standard a.c. power, allowing you to use your computer's standard power adapter/charger in the field. Power inverters do use a substantial amount of battery power, so go with the 12V power adapter if it is available for your computer.

Required Cables

Next, you will need a cable. The same cable works for all NexStar models. The basic requirement is a cable that mates the correct pins from a DB-9 connector (found on the computer) to an RJ-22 connector (found on the bottom of the NexStar hand control). The RJ-22 connector is the same as found on the handset of a typical telephone – it is also referred to as a 4P4C connector. This is not the same as the RJ-11/12 used to connect a telephone to the wall jack. The correct pin connections for the standard NexStar cable are shown in Figure 10.1. Note that this drawing represents the end of the cable that connects to the bottom of the NexStar hand control. Figure 10.2 shows the correct connections for the less commonly used DB-25 serial port found on some older computers.

The cables required for connecting to a palmtop are a bit more complicated. The standard serial port on a laptop or desktop computer uses a DB-9 male connector. Thus, the standard NexStar cable provides a DB-9 female connector. The serial sync cable for a palmtop provides a DB-9 female connector and is wired with pins 2 and 3 backwards of the serial port on a computer. Thus, to use a standard NexStar cable with a palmtop serial sync cable, you will potentially need two adapters: a null modem adapter (always required) and a gender changer if the null modem adapter does not connector directly between the sync cable and the NexStar cable.

Most telescope equipment vendors stock the standard NexStar cable for about $25. Null modem and gender adapters are available from computer parts stores or from the Software Bisque web site (http://www.bisque.com). I highly recommend the cables sold by

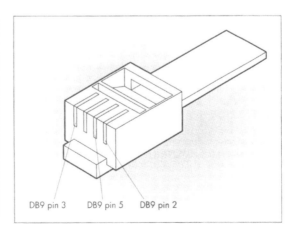

Figure 10.1. Pin connections for DB-9 serial ports.

DB9 pin 3   DB9 pin 5   DB9 pin 2

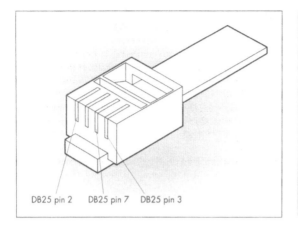

DB25 pin 2   DB25 pin 7   DB25 pin 3

Figure 10.2. Pin connections for DB-25 serial ports.

Astro Gadgets (http://nexstar.astrogadgets.com). The cost is much lower than from other suppliers and the quality is high. They also sell NexStar cables designed for palmtop computers that do not require a null modem adapter or a gender changer – these are easily worth three times the modest cost they ask.

If you decide to build a cable, here are some tips. Buy a telephone handset cable, the coiled type that goes from the handset to the telephone. A standard telephone cable does not have the correct connector required for the NexStar hand control. Be sure it has four colored wires visible at both ends; some cables only have two. Remove one of the RJ plugs and attach a DB-9 female connector matching the pins shown in Figure 10.1. For a palmtop sync cable, attach a DB-9 male connector and reverse the connection of pins 2 and 3. It is best to make the DB-9 end by crimping an RJ-11 connector (a standard phone connector) on one end of the cable and using a DB-9 to RJ-11/12 adapter, which can be found at computer parts stores or A2Z Cables (http://www.a2zcables.com). If all of this seems too complicated, better buy a pre-made cable.

Most NexStar cables are 12 to 15 feet long. Some have experimented with longer lengths and success has been reported with distances of 50 feet. Another option is to go wireless. There are RS-232 wireless adapters available, but the cost is significant. The market is small for such devices and that market is shrinking due to the replacement of RS-232 with newer connection interfaces. My most recent searches on the Internet turned up several sources in the $300-and-up price range.

Software to Control NexStar Telescopes

There are a wide variety of software packages available for controlling your NexStar telescope. Prices range from free to a few hundred dollars. While many of the packages are similar in function, some are very unique. I recommend you spend some time researching manufacturers' web sites to determine which software package or packages best suit your needs.

When researching software for use with your NexStar telescope, you should be aware of the various models. Not all programs that claim NexStar compatibility work with all models. Table 10.1 summarizes the various models:

Table 10.1. NexStar models summarized

Model	Description
Original GT	The original GT (computerized GoTo) hand control for the NexStar 60/80/114/4 telescopes, as well as all Tasco StarGuide telescopes.
New GT	The upgraded GT hand control for the NexStar 60/80/114/4 telescopes. This hand control is included with all little NexStars manufactured after December 2001. See the sidebar "Old Versus New GT Hand Control" in Chapter 3 to determine whether you have the original or new GT hand control
NexStar 5/8	The original NexStar 5 and 8 models. These models were discontinued in the summer of 2002 and replaced with the NexStar 5i and 8i
NexStar 5i/8i	The new NexStar 5i and 8i telescopes. Only the optional computerized hand control allows PC control.
NexStar GPS	The NexStar 8/11 GPS telescopes.

The New GT, NexStar 5i/8i, and NexStar GPS use the same basic control commands. Thus, a program that states compatibility with one will generally work with the others, although some features in the program might not work with all models. For example, a program with manual slew buttons that emulate the direction buttons on the NexStar hand control will work with the NexStar 5i/8i, GPS, and New GT models, but that specific feature (manual slew buttons) will not work with the New GT hand control. The Old GT and NexStar 5/8 are each unique and not compatible with the others or each other – if you own one of these you should definitely ask if the software publisher does not clearly state compatibility with these older models.

Appendix C is a list of all NexStar-compatible programs known to me at the time of this writing. Note that the free version of TheSky (Level 1) that ships with some NexStar telescopes does **not** include telescope control. Refer to Appendix C for versions of TheSky that do provide telescope control.

One of the packages mentioned in Appendix C is the ASCOM driver set. The ASCOM driver is not actually a complete software package but rather is used by other programs to interface with the telescopes that it supports. This allows a single telescope driver to be optimized for each model of telescope and authors of astronomy software are relieved of the task of creating and maintaining drivers in their programs as new telescopes are introduced. In fact, an updated ASCOM driver can be installed on your computer without requiring changes to the astronomy software that uses it.

There are a variety of commands, sometimes referred to as RS-232 commands, that NexStar telescopes respond to. The basic commands direct the NexStar to GoTo a set of RA–Dec coordinates and query the scope for the current RA–Dec it is pointed towards. All models of NexStar telescopes respond to these commands and most PC and palmtop software makes use of them. Additional commands are available to allow direct motion control, in essence emulating the four arrow buttons on the hand control. Only the NexStar 8/11 GPS and 5i/8i provide this capability. Direct motion control is needed to provide on-screen slew buttons, smooth satellite tracking, and software-based autoguiding for astrophotography. Slew buttons are now available in some software and the ASCOM driver. Satellite Tracker provides smooth tracking of satellites and the International Space Station. Software-based autoguiding is still under development at the time of this writing.

Putting It All Together

There is no preferred order, but you will need to connect your PC or palmtop computer to the bottom of the hand control, power up the computer, and perform an alignment of the telescope. If you are testing indoors, use Quick Align or if your model of NexStar doesn't support that method, use Auto Align and accept the default locations for the two alignment stars. On the NexStar 5/8 (original model) and the little NexStars with the new GT hand control, press the Menu button, scroll to RS-232, and press the Enter button.

Start your astronomy software and your computer should be ready to communicate with your NexStar. Some programs have a menu item or toolbar button to establish communication while other programs are ready to talk immediately after starting; refer to the documentation for your program if this is not readily apparent. A good test for communication is to direct your program to retrieve the telescope's coordinates. Other than the original GT hand control, the RA and Dec returned should match the coordinates reported by the Get RA–Dec command on the hand control Menu button. The next test would be to use the program to direct the scope to GoTo an object. After the scope slews to the requested coordinates, check with the hand control's Get RA–Dec command. If the coordinates match within a few arc minutes, then all is working properly. If the telescope doesn't move at all, check the LCD panel on the hand control for a slew limit warning - the object you requested might be outside your established Slew Limits.

If things don't seem to be working as expected, refer to the next section for possible solutions.

Common Problems with PC to Scope Communication

Most NexStar owners experience no problems controlling their telescope with a computer. However, computers are sometimes finicky and there are several potential problems you may come across. First, you will want to be sure you are using a program that is compatible with your model of NexStar telescope. If you are not sure about your program and are using a computer with Microsoft Windows, I would recommend you download the freely available NexStar Observer List (NSOL) from my web site (http://www.NexStarSite.com). NSOL uses the most straightforward method possible for communicating via the serial port, so if you get NSOL working with your telescope then, other programs should work just fine.

Following are some tips to get your PC and scope communicating. Note that some of these tips are related to Microsoft Windows and may not apply directly to your situation.

- The telescope must be aligned before it will accept commands. To test indoors, simply perform a Quick Align or an Auto-Align, accepting the locations pointed out for the two stars.

- For some models of NexStar (N5/8 and the new version of the N60/80/114/4), you must enter "RS-232" mode from the hand control menu before the scope will accept commands.

- A faulty cable might be the problem. One way to test the cable is to install NSOL on a standard desktop PC and try controlling the scope from there (remember the indoors alignment from the first item above). Desktop PCs do not generally have much problem with conflicting comm ports, *except* in the case of software to synchronize a PDA such as a Palm Pilot.

- With homemade cables a common problem is incorrect pin connections. Check the cable against Figure 10.1, again noting that this figure shows the end of the cable that connects to the hand control.

- Check the port on the bottom of the hand control for bent pins – these can easily be straightened with a small screwdriver.

- Be sure there are no other programs using the serial port. For example, installing Palm Pilot software on a computer is a sure way to disable the serial port for access by other programs like NSOL. If there is a little icon for synchronizing in the system tray (the section in the lower right corner of the screen with the current time), then it is possible that a synchronizing program has taken control of the serial port. You can generally close such programs by right-clicking the icon in the system tray and choosing Exit. If you no longer use that program, try uninstalling it.

- Try other "Comm Port" settings in the program. In NSOL, go to the Tools menu, choose Setup and try settings other than Com1. If this resolves your problem, be sure to record your setting for use in other programs.

- If you are using Microsoft Windows, go to Control Panel (Start menu, Settings) and open the System icon. Click on the Device Manager and look for "Ports (COM & LPT)". There should be a "+" mark next to it and when you click the +, Device Manager will expand to show the actual ports. At least one of the ports should be labeled "Communications Port (Com1)" or perhaps Com2. If the little icon next to this entry has an X through it, you must double-click the icon and try to determine why the comm port is not operating.

- Another potential conflict, particularly with a laptop computer, is an infrared (IrDA) port. In Microsoft Windows you can find it listed in the Device Manager as well. If you are not using the infrared port, double-click the icon (not the one with the + mark, but rather the one below it) and check the box to "Disable in this hardware profile." Other methods for disabling the infrared port are generally explained in the manual for your computer.

- There is a known problem with the serial port on some newer computers – notably Dell and Toshiba computers. Sometimes they will communicate with the scope, sometimes they just won't. If you have tried all of the above and get no communication or intermittent communication, purchase a USB-to-serial-port adapter and use that rather than the built-in serial port. Also, check the computer manufacturer's web site, a BIOS update that fixes the problem might be available.

- Most other potential solutions require investigating and possibly changing the settings for serial and/or IrDA ports in the computer's setup (BIOS settings). This requires reasonable experience in configuring computers and is not recommended unless you are already familiar with such procedures.

Computers are fairly complex, and there are other possible sources of trouble, but hopefully one of these tips will resolve any issues you have.

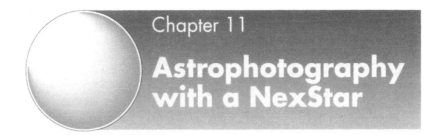

Chapter 11

Astrophotography with a NexStar

There is something alluring about using your own equipment to produce a photograph of a celestial object. Hanging in your home or office, such photos never fail to draw attention and comments. When asked if you got the image from a magazine, you can imagine the pride experienced when you say, "No, I took that myself."

What you may not be able to imagine is the hours of hard, often agonizing work that you will endure to create a photograph you can proudly display. For every acceptable image produced, many more will be thrown out. But certainly that makes the good ones all that much more special.

A true introduction to astrophotography is a book unto itself. In this chapter I will try to give you the big picture, as well as point out NexStar-specific issues. For additional information, I recommend:

- *Astrophotography for the Amateur* by Michael A. Covington – an excellent introductory book.
- *The New CCD Astronomy: How to Capture the Stars With a CCD Camera in Your Own Backyard* by Ron Wodaski – the definitive book on CCD astronomy for the amateur astronomer.
- http://groups.yahoo.com/group/digital_astro – a discussion group on astrophotography using digital cameras.
- http://groups.yahoo.com/group/videoastro – a discussion group on astrophotography using video cameras.
- The QCUIAG web site – http://www.astrabio.demon.co.uk/QCUIAG – a great collection of information and tools for using web cams and other video cameras for astrophotography.

I will give you one specific tip – you **must** keep accurate notes. Although there are many guidelines that will help you to produce good images, there is a lot of trial and error involved. I recommend you keep a logbook for your imaging efforts, separate from your observing log. For every exposure you should record the object being imaged (include the phase of the Moon if that was your target), the equipment used, the resulting focal ratio, the camera settings, the exposure time, the type of film (if applicable), and the seeing conditions. This historical reference will be invaluable as you progress.

Astrophotography Basics

Camera Types

A wide variety of cameras are useful for astrophotography – chances are you may already have a suitable camera to get you started.

35mm SLR Cameras The traditional camera for astrophotography is a 35mm SLR with interchangeable lenses. Many people have just such a camera tucked away in the closet. Up front though, you should know that film requires longer exposures than all other methods we will discuss and thus requires an accurately tracking telescope mount with autoguide capability as a very attractive option. The only NexStar telescopes that match this description are the NexStar 8/11 GPS and the NexStar 5i/8i. Additionally, the longer exposures will require the scope to be polar-aligned on a wedge.

When considering a 35mm SLR, note that some are better suited to astrophotography than others. The most critical feature is the ability to hold the shutter open for long periods of time to allow the faint light of deep sky objects to accumulate on the film. Look for a "B" shutter speed setting. With this setting the shutter will stay open as long as the button is pressed. To reduce shake and hold the shutter button in place, you will need a locking shutter release cable. Another nice feature is a camera that manually holds the shutter open. Most 35mm SLR cameras rely on the battery to hold the shutter, a serious disadvantage, especially in cold weather when batteries quickly loose power.

Focusing on faint objects is surprisingly difficult with most cameras. Some 35mm SLR cameras have interchangeable focusing screens. Focusing screens that allow maximum light to pass through aid in focusing. Another useful feature is a viewfinder magnifier to further ease focusing. These accessories are available at fully stocked camera stores.

When selecting film and total exposure times, a little experimentation is in order. There are many variables involved, primary being film speed, shutter speed, focal length of the telescope, prime focus or eyepiece projection, and the inherent brightness of the object being photographed. A full description is well beyond the intended scope of this book, so I would refer you to any of a good number of books on astrophotography. Your local library, bookstore, or the Internet would be places to start. In particular I recommend *Astrophotography for the Amateur* by Michael A. Covington and *The Backyard Astronomer's Guide* by Terence Dickinson and Alan Dyer.

Digital Cameras A much easier way to get your astrophotography feet wet is with a digital camera (Figure 11.1). While not well suited for fainter objects – even the best commercial digital cameras can capture images of only the brightest deep sky objects – wonderful images of the planets and the Moon are being created regularly with digital cameras. A great resource for information as well as getting your questions answered is the digital camera astrophotography discussion group on Yahoo: http://groups.yahoo.com/group/digital_astro.

Additional processing of images is generally accomplished on a personal computer (PC); Windows or Macintosh computers are commonly used. It is common to capture several images in rapid succession and then electronically "stack" them in the PC. The resulting image has improved detail with reduced noise. To take advantage of this technique, a remote control for the camera is a great accessory. Using the remote, you can easily snap multiple images without shaking the camera.

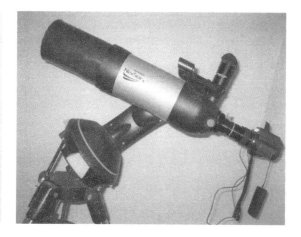

Figure 11.1. NexStar 4 on a wedge. Attached to the rear port: LAR, SCT visual back, eyepiece, digital camera adapter, and digital camera. Photo courtesy Jordi Sese.

Video Cameras

Traditional video cameras (video tape camcorders) and PC-based video cameras (often called web cams) are another way to capture the brighter objects in the sky. With the addition of a small video monitor or TV, it is also a great way to share these objects with a group in real time. Some of the better astronomy-specific video cameras will even let you capture and display relatively faint deep sky objects in real time.

At some point, if you intend to produce printed photographs or post your images on the Internet, a computer will be required. With some video cameras, you will be able to transfer video files to your computer via either a USB or an IEEE 1394 port. For other video cameras you will need to purchase a video capture device. Converting a video file to a still image requires additional processing with a variety of PC-based software available. In essence, a video file is hundreds of snapshots that can be stacked for greatly improved detail.

In the past those interested in video astronomy were forced to work with less than optimal web cams, usually performing permanent modifications to make them more suitable to the task. Still, these modified web cams were only capable of low-resolution images of the brighter celestial objects. Today, you will find a wide variety of video cameras specifically manufactured for astronomy. Lower-priced models, generally no more than about $200, are available from SAC Imaging (http://www.sac-imaging.com), Sky-AstronomyQuest.com (http://www.astronomyquest.com), Adirondack Video Astronomy (www.astrovid.com), and others. These lower-priced models are still only suitable for brighter objects, but they are ready out of the box.

A very exciting class of video cameras for astronomy is based on newer, super-sensitive CCD chips. Adirondack Video Astronomy (http://www.astrovid.com) offers the Astrovid StellaCam EX, ITE Telescopes (http://www.itetelescopes.com) has the DeepSky Pro B&W and the slightly more sensitive DeepSky PRO EX B&W, and for a highly sensitive color video astronomy camera consider ITE's ColorEye PRO. All of these cameras integrate multiple video frames on the fly, allowing deep sky objects to be displayed on a video monitor or captured on videotape or a PC.

Note that many astronomy-specific video cameras are black-and-white only. While this is not a disadvantage for the Moon and most of the deep sky objects they can capture, you will most likely want color for Jupiter, Saturn, Mars, and some nebulae. As a tradeoff, the better black-and-white video cameras are much more light-sensitive than color video cameras, allowing the black-and-white models to show much fainter objects. An alternative is to use a black-and-white camera to capture three video clips through different colored filters and then digitally combine the results.

To get started visit the QCUIAG web site – http://www.astrabio.demon.co.uk/QCUIAG. To ask questions and for feedback on your efforts, join the Yahoo video astronomy discussion group- http://groups.yahoo.com/group/videoastro.

CCD Astronomy Cameras (Figure 11.2) For long-exposure images of faint objects, you really have two choices: film cameras or CCD astronomy cameras – generally just referred to as CCD cameras. CCD cameras enjoy several advantages over film. First, they require shorter exposure times to capture faint objects. Another advantage is that CCD cameras do not suffer from the characteristic of film known as reciprocity failure. During longer exposures, film steadily looses its ability to record additional light and eventually fails altogether. CCD cameras continue to capture light in a steady, linear fashion throughout the exposure. A final clear advantage is the nearly instant feedback you receive with a CCD camera. You can view the captured images on a computer screen immediately without the need to process a roll of film.

Film does still have some advantages. Most CCD chips are very small and thus cannot capture large sections of the sky. For larger objects this may require multiple images to be tiled together to create a complete composite of the subject. This is steadily changing as larger astronomy-capable CCD chips are now being released on the market, but they are very expensive. Another area where film still offers an advantage is the realm of color imaging. Most CCD cameras are black-and-white. To produce color images it is necessary to capture at least three images, each through a different colored filter, and then combine them digitally on a computer. Typically this is accomplished with a filter wheel that holds all required filters and rotates them into position as necessary. Some filter wheels are even controlled automatically by a computer. One final advantage is overall cost. A 35mm film camera is cheaper than even an introductory level CCD camera. And most CCD cameras will require a computer alongside the telescope when you are capturing images.

Figure 11.2. SBIG ST-237A CCD camera on a NexStar 11 GPS. Photo courtesy Jeff Richards.

In some ways, the term "CCD camera" is ambiguous. The heart of such a camera is a CCD (charge-coupled device) chip. Most digital cameras and video cameras also capture their images with a CCD chip. But CCD cameras are designed for exposures of many minutes while digital cameras seldom allow exposures of more than about one minute. Video cameras are designed to capture multiple exposures *each* minute. Longer exposures on digital cameras are not practical due to noise buildup that shows itself as light-colored flecks throughout the image. This noise can be reduced if the CCD chip is cooled.

Since the CCD chip is mounted in the dark insides of the camera, altering a camera for cooling is generally impractical (you will find that longer exposures are possible with a digital camera during very cold weather). CCD cameras incorporate CCD chips with improved light sensitivity *and* such cameras employ active cooling. They are, in effect, "air-conditioned". This is why they are capable of the longer exposures required to capture the faintest objects.

The most popular models of CCD cameras come from Santa Barbara Instrument Group (SBIG) (http://www.sbig.com), Apogee Instruments (http://www.ccd.com), and Starlight Xpress (http://www.starlight-xpress.co.uk). Starlight Xpress has the distinction of producing the only full-color CCD astronomy cameras. Meade Instruments (http://www.meade.com) also manufacturers CCD cameras, but the comparably priced models from the big three are recommended.

SAC Imaging has recently introduced two new models, the SAC7 and SAC8, providing low-cost, cooled CCD cameras somewhere between traditional CCD camera and video camera technologies. The SAC7 series are full-color; the SAC8 is black-and-white. They are relatively low-cost and provide an affordable entry into long-exposure astrophotography.

Note that most CCD cameras use either the parallel port or the serial port to transfer images to a computer. Newer CCD cameras use a USB interface for much faster image transfer – a highly recommended feature.

Capturing the images and processing them after transferring them to a computer will be required. The software that comes with the camera will generally suffice to get you started, but most serious imagers elect to use alternate software to further automate the image processing and capture process. Popular software packages such as MaxIm DL from Cyanogen Imaging Products or CCDSoft from Software Bisque (http://www.bisque.com) can greatly reduce the effort required to get great images.

Be careful when choosing a CCD camera for your telescope. The objects you will image and the focal length of your telescope will dictate the cameras best suited for the task. Again, this topic is beyond the scope of this book, so prior to spending thousands of dollars on a CCD camera, the book *The New CCD Astronomy: How to Capture the Stars With a CCD Camera in Your Own Backyard* by Ron Wodaski is required reading.

Mounting the Camera

There are several different ways to use a camera to photograph the night sky. To some extent the camera will dictate which techniques you can apply.

Piggyback Any camera with its own lens can be mounted piggyback on larger scopes. The camera is mounted on top of the scope, with a specific piggyback bracket or generic mounting rails (see Chapter 8 for more details). In addition to a mount, you should also consider a slow-motion adapter such as are available from Orion, and Baader Planetarium (Baader calls theirs "Witty 1"), or most camera shops. These adapters attach between the mount and the camera, allowing a fine level of control over the pointing of the camera. This greatly improves your ability to compose a shot while using the main telescope to guide on any bright star in the area.

Exposures of several minutes are generally required to capture faint nebulae and star fields. 35mm SLR and CCD cameras are commonly used, although some have experimented with digital cameras. A camera with interchangeable lenses is desirable to allow various magnifications and fields of view to be captured. Piggyback astrophotography almost always requires polar-alignment on a wedge due to the longer exposures.

Afocal With the afocal method, a camera, with its lens in place, is mated to the eyepiece of the telescope. This is the best (in most cases *only*) method suitable for digital cameras, video camcorders, and standard web cams. In fact, afocal astrophotography was seldom practiced until digital cameras became widely available. It is possible to simply hold the camera up to the eyepiece, but much better results are possible with special adapters to hold the camera securely to the eyepiece. The most critical key to good afocal imaging is positioning the lenses of the camera very closely to the first lens of the eyepiece. Indeed, they should almost touch. To insure such close positioning, the proper eyepiece and mounting adapter must be carefully selected.

Prime Focus When imaging at prime focus, the camera lens is removed and the camera is mounted directly in place of the telescope eyepiece. The telescope becomes the camera's lens. This method is commonly used with 35mm SLR cameras, CCD cameras, and video cameras designed for astronomy. Since the camera is placed where the eyepiece is normally located, it is common to use a "flip mirror" or off-axis guider to allow both a camera and an eyepiece to be used. The diagonal is replaced with the flip mirror or guider. The camera is mounted at the rear and an eyepiece is mounted in the tube coming straight up from the side of the flip mirror or guider. If using an off-axis guider, after the image is centered, an autoguider will often be used in place of the eyepiece as shown in Figure 11.3.

Due to relatively long exposure times, prime focus astrophotography requires very accurate tracking. Long exposures also require polar alignment on a wedge, and guiding (manual or auto) is generally needed. The NexStar 4 and all of the SCT models are capable of meeting these requirements, although only the NexStar 5i/8i and NexStar 8/11 GPS allow autoguiding (discussed later in this chapter).

Figure 11.3. Off-axis guider with CCD autoguider and 35mm camera. Photo courtesy Andrew Riehl.

Eyepiece Projection While prime focus astrophotography produces stunning images of extended objects, the added magnification required for smaller subjects, such as the planets and close-ups of the Moon, is provided with the eyepiece projection technique. Eyepiece projection is similar to afocal except the lens is removed from the camera and the camera is mounted above the telescope's eyepiece. Just as with visual observation, different eyepieces can be used to vary magnification. Eyepiece projection places even more severe demands on tracking and guiding than prime focus photography.

Focusing

Often we find it difficult to achieve good focus even when visually observing at the eyepiece. Focus during imaging is even more critical to capturing the best possible images. In fact, even the slightest deviation from correct focus becomes painfully obvious during longer exposures. To make matters worse, it is **much** more difficult to focus the telescope with a camera mounted.

As discussed earlier, you may find it necessary to replace the focusing screen of a 35mm SLR camera to allow enough light through the viewfinder for accurate focusing. Also, a viewfinder magnifier aids the process. With most CCD cameras you must adjust the focus, wait for the updated image to transfer to the computer screen and repeat until you achieve a good focus. Focusing with a digital camera requires a camera with a small LCD preview panel; otherwise you are simply "shooting in the dark." The small viewscreen on most camcorders allows fairly accurate focusing, but for an easier time of it, connect the camera to a small television or video monitor. The same suggestion can be used with many web cams and even some digital cameras.

While you will likely find these techniques sufficient to focus on the planets and the Moon, focusing on stars and other deep sky objects is truly difficult. Many find the Hartmann mask indispensable, especially when imaging at prime focus. The Hartmann mask is simply a cardboard cover that fits over the front of the telescope. Two or three holes (refer to Figure 11.4) are cut into the cardboard and focus is achieved when the light from all the holes merges into a tight pattern. Kwik Focus from Kendrick Astro Instruments is a ready-made Hartmann mask, or you can easily make your own.

A newer technique said to provide even better focus utilizes two bungee cords (thick elastic cords) stretched across the front of the scope at right angles. The cords form an "X"

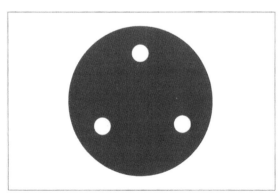

Figure 11.4.
Hartmann mask.

centered on the front of the scope. When pointed at a bright star, the cords will cause diffraction spikes easily visible in the camera. Focus until the spikes are as small as possible.

And finally, focusing is considered so crucial that computer software has been designed to analyze the image from CCD cameras and drive motorized focusers to automatically provide the best image. MaxIm DL from Cyanogen and PCFocus/FocusAide from Adirondack Video Astronomy are two examples.

Accurately Tracking the Sky

Exposures longer than just a few seconds require a mount that accurately tracks the sky. Additionally, exposures longer than about a minute require a polar-aligned mount. For NexStar telescopes, polar alignment requires a wedge to tilt the scope such that the azimuth axis (a line coming straight up from the center of the base) is in line with the axis of the Earth. Another way of saying this is that the fork arms are pointed at the north or south celestial pole.

As noted in Chapter 8, the NexStar 4, NexStar 5/5i/8/8i, and NexStar 8/11 GPS can all be easily mounted on a wedge. Figure 11.5 shows a NexStar 11 GPS on a wedge. Mounting the other NexStars on a wedge would require that you build a wedge or adapter yourself. After mounting on a wedge, use the alignment methods discussed in Chapter 4 to get a fairly

Figure 11.5. NexStar 11 GPS polar-aligned on a wedge. The fork arms are pointed towards the north celestial pole. Photo courtesy Bob Berta.

accurate polar alignment. For exposures longer than a few minutes, you will likely need to refine the polar alignment with a method like "drift alignment".

Even with a very good polar alignment the gears in most mounts are not accurate enough to allow unguided images longer than just a few minutes. The NexStar 8/11 GPS include a feature known as Permanent Periodic Error Correction (PPEC) to improve the maximum possible duration of unguided images. PPEC, described in Chapter 5, allows longer unguided images, although the maximum time is dependent upon the focal length of your configuration, the subject's placement in the sky, and other factors.

Longer exposures will usually require that the telescope be guided. Guiding can be performed by manually keeping a star centered in a reticle eyepiece or by an autoguider connected to the NexStar 5i/8i and NexStar 8/11 GPS models' Auto Guide port. The autoguider contains a CCD chip and attempts to keep a star located on the same pixel (picture element) for the duration of the exposure. Guiding requires either a piggyback-mounted guide scope (a small refractor telescope) or an off-axis guider (shown earlier in Figure 11.3). The reticle eyepiece or autoguider is mounted in either the guide scope or the off-axis guider. Some CCD cameras and astronomy-specific video cameras can be used to both guide and capture an image. The camera alternates between the two tasks and thus an off-axis guider is not needed for a separate autoguider.

Digital Image Processing

Images captured with a digital or CCD camera are digital in nature. Images captured with a video camera must be converted to digital (as discussed earlier) if you wish to make prints or post the image on the Internet. While it is possible that the raw images will be of suitable quality for presentation, most astrophotographers process their images in various PC-based software to bring out the best possible appearance.

The most popular processing software for the final image is Adobe Photoshop, although other packages can also be used. Advanced techniques such as dark frame subtraction, unsharp mask, and color balancing are used to improve contrast and overall appearance. Photoshop is also useful to combine the black-and-white CCD images taken through color filters to produce a final color image.

As discussed earlier, multiple video frames and digital camera exposures are often "stacked" (combined) to produce a final image with greater detail and contrast. Two useful programs for stacking are AstroStack (http://www.astrostack.com) and Registax (http://aberrator.astronomy.net/registax) both free for download.

Astrophotography with NexStar Telescopes

Digital and Video Cameras

All NexStar models are capable of fine images using a digital camera in an afocal manner. All NexStar scopes can be used successfully for video imaging as well. In both cases the biggest challenge might be balancing the added load of the camera, especially when using the NexStar 60/80/114/4 scopes. The added weight can cause terrible tracking and in extreme cases the scope may not be able to track at all. The best solution is to add weight to offset the load. Ankle weights used when exercising are a good suggestion, but many possible solutions exist if you use a little creativity.

When mounted in the altitude–azimuth mode, the tracking of the earlier NexStars – the 60/80/114/4/5/8 models – is not usually accurate enough for digital camera exposures longer than just a few seconds. In fact, the simple motion of tracking can cause blurred images in the shortest exposures. If you experience this problem, after focusing with the camera mounted turn tracking off (on the Menu button) when taking images. GoTo still works and you can manually center your object and snap shots after any vibrations subside.

Long-Exposure CCD and Film Imaging

The NexStar 8/11 GPS were designed from the ground up with this in mind. Their rock-solid design, PPEC, and autoguide capability provide a great platform for long-exposure imaging. They also offer the versatility of multiple focal ratios when using focal reducers and Fastar capability. The NexStar 5i/8i can let you get your feet wet, although the single fork arm and spur gears do not offer the best stability and tracking. To their advantage the NexStar 5i/8i can also use focal reducers and offer autoguide capability as well. The NexStar 4/5/8 can be used on a wedge for longer exposures, but you will need to manually guide any exposures longer than a minute.

Correctly balancing the scope is the key to the precise guiding and tracking required for long-exposure astrophotography. For objects in the eastern sky, balance the scope to be a bit front heavy in the RA (azimuth) axis. For objects in the western sky, change the balance to back heavy. This is when a sliding weight set, available for the NexStar 8/11 GPS and discussed in Chapter 8, proves very useful. The key is to keep the RA (azimuth) motor driving the gears against the weight. An alternate method is to use weights on the lower fork arm, the one that is being lifted by the rotation of the RA axis.

In relationship to the Dec (altitude) axis the scope always needs to be a little unbalanced, front- or back-heavy, when mounted on a wedge. This prevents backlash from causing problems with guiding and tracking on that axis. If the scope is perfectly balanced in Dec, it can "float" from one side of the gear play to the other. For some orientations of the optical tube, it is necessary to place weights along the bottom **and** top of the telescope to achieve the correct balance on both axes.

Occasionally you will choose to image an object that is very faint; some may actually be too faint to see visually. The film or CCD chip will be able to record such faint images – but only if you get them centered in the frame. For added GoTo precision you might find it useful to use the Re-Alignment feature described in Chapter 4. Realign on a star near to your target. Then when you GoTo that faint object it should be located near the center of the CCD chip or film frame.

The Auto Guide port on the NexStar 5i/8i and NexStar 8/11 GPS is designed to be compatible with the SBIG ST4 standard. The ST4 operates like four simple switches, one for each of the four slew directions. Most autoguiders use electronic switches and require a relay box as an interface between the autoguider and the telescope. Some have successfully used their autoguider without a relay box, but I would recommend the relay box. In addition to providing the required interface, the relay box also electrically isolates the autoguider/camera from the telescope electronics. The motors in the scope are a potential source of electrical noise that can spoil images in some cameras.

Figure 11.6.
Standard f/10
configuration compared
to Fastar f/2
configuration. Drawings
courtesy Celestron.

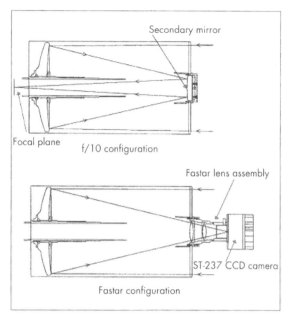

Celestron's Fastar System

The NexStar 8/11 GPS telescopes incorporate a unique capability known as Fastar. When imaging, faster focal ratios (f/5 for example is faster than f/10) allow for shorter exposure times and capture a wider section of sky in any given frame size. As we will discuss in the next section, focal reducers are typically used to provide a faster focal ratio, but Fastar exceeds what is possible with a focal reducer.

The NexStar 8/11 GPS telescopes have removable secondary mirror assemblies. A threaded lockring holds the secondary in place, yet allows easy removal. After removing the secondary mirror, a Fastar lens assembly is mounted in its place. In turn, a CCD camera is mounted on the Fastar lens. The result is a focal ratio of approximately f/2 (Figure 11.6).

The standard focal ratio of the NexStar 8/11 GPS scopes is f/10. Images taken at f/2 require exposures 25 times shorter than f/10. A faint nebula requiring three 20-minute exposures through color filters can now be completed with three exposures of just 1 minute each. Fastar is not suitable for every object, but it is a wonderful way to get started with many deep sky objects. An added advantage is that most objects can be imaged without guiding due to the short exposure times.

One caveat exists – the only camera supported for all Fastar models is Santa Barbara Instrument Group's ST-237. The ST-237 is quite affordable – in fact it is one of the least expensive cameras SBIG makes – and offers advanced features such as Track and Accumulate mode (autoguiding and imaging with a single camera) and an optional integrated filter wheel for color imaging. With the correct adapters some Fastar lens assemblies are compatible with the MX5C and MX7C cameras from Starlight Xpress. It is best to ask before you buy to insure a specific camera is compatible with Fastar.

Celestron makes a Fastar lens (model 94180) for the NexStar 8 GPS. Starizona (http://www.starizona.com) makes and sells its own Fastar-compatible lens assemblies

Figure 11.7.
HyperStar and ST-237
on the NexStar 11
GPS. Photo courtesy
Starizona.

under the name of HyperStar. They have models for both the NexStar 8 GPS and the
11 GPS (Figure 11.7). HyperStar lens assemblies provide an even faster focal ratio
than Celestron's offerings.

Adapters for Astrophotography

The various methods of photography discussed in this chapter require adapters for attaching the camera to the telescope or reducing the focal length to shorten exposure time.

Prime Focus and Eyepiece Projection Adapters The most
common method of attaching 35mm and some CCD cameras to a telescope is with the use
of a standardized thread referred to as a T-Mount (Baader Planetarium calls it T-2). A T-
Ring made for the specific camera is required to mate the camera to the T-Mount. While
the end of the black focuser tube of the NexStar 80 is threaded as a T-Mount, all other
NexStar models require a T-Adapter to provide the T-Mount threading that then mates to
the camera's T-Ring.

Standard T-Adapters, sometimes called universal adapters, are available for insertion in
a 1.25in focuser tube or visual back. Schmidt–Cassegrain telescope (SCT) T-Adapters
thread to the rear of all NexStar SCT models and are preferred as they offer a wider
opening that illuminates more of a 35mm film frame. For the NexStar 4, good choices are
Orion's "T-Adapter for the C90 and Meade ETX telescope", Baader Planetarium's "C90/N4
T-Adapter", or Meade's "T-Adapter for ETX". These standard T-Adapters allow prime
focus photography, providing the widest possible view (lowest magnification) during
imaging.

When using a CCD or film camera to image smaller objects such as the planets and the
Moon, we need to pump up the magnification. For this, we turn to eyepiece projection
adapters. As with T-Adapters, eyepiece projection adapters are made to fit into a 1.25in
focuser tube or thread directly to the rear of an SCT. An eyepiece is slid into the adapter to
provide the needed magnification. The top of the adapter is threaded as a T-Mount, pro-

viding the needed connection for the camera. Eyepiece projection adapters are available as either standard or variable adapters. Standard adapters are a set length and the only way to change the magnification is to change the eyepiece in use. Variable adapters are adjustable in length, allowing changes in magnification by either changing the length or by changing the eyepiece.

Celestron, Baader Planetarium, Meade, and Orion manufacture a variety of T-Mount and eyepiece projection adapters. The adapters from Baader Planetarium are especially notable due to their unique designs that allow great flexibility in mixing and matching adapters to get the required configuration.

Digital and Video Camera Adapters

With the exception of astronomy video cameras designed to fit in the place of an eyepiece, all other video cameras and digital cameras are used in the afocal method. Better digital cameras and camcorders have threads in the front of their lens assembly for the use of filters. When this is the case, a threaded adapter is the best choice for a close mating of the camera and the eyepiece. For cameras without filter threads, a large bracket is necessary to hold the camera near the eyepiece. In practice, it is extremely difficult to adjust such brackets to keep the camera lens closely mated to the eyepiece. To provide easy focusing and reduce vignetting (darkening around the edges of the image), it is crucial that the camera lens be as close as possible to the eyepiece and that the camera sit squarely above the eyepiece.

ScopeTronix (http://www.scopetronix.com) manufactures adapters for a wide variety of cameras. Most of their adapters require the use of an eyepiece that has a removable rubber eyecup to allow three screws to secure the adapter to the eyepiece. If you would rather not mar your eyepieces and for additional flexibility, Baader Planetarium's OPFA system (http://www.alpineastro.com and http://www.baader-planetarium.de) holds eyepieces internally in the same fashion as an eyepiece projection adapter. Threaded adapter rings mount on top of the OPFA adapter for connecting to your camera.

Off-Axis Guiders and Flip Mirrors

As mentioned earlier, with a CCD or other camera mounted in prime focus or eyepiece projection configurations, a flip mirror or off-axis guider is useful to insure the image is centered in the field. An off-axis guider is also convenient to manually guide an image without use of a piggyback guide scope. Celestron, Meade, the now defunct Lumicon, and Van Slyke Engineering make off-axis guiders and flip mirrors. Most prefer the Lumicon and Van Slyke models. It should also be noted that the NexStar 4 has an internal flip mirror and a rear port for mounting a camera.

Focal Reducers

A shorter focal length means shorter exposure times. Also, shorter focal lengths are more forgiving of tracking errors caused by the mount or poor polar alignment. In addition to the Fastar lens assembly, there are other options to reduce the focal length of a Schmidt–Cassegrain telescope. SCT focal reducers thread directly to the rear of 5- and 8-inch SCTs (Figure 11.8). On the NexStar 11, the focal reducer threads to the large adapter ring. A standard visual back (1.25in or 2in) or camera adapter is then threaded to the focal reducer. Celestron and Meade make f/6.3 focal reducers and Meade also makes an f/3.3 focal reducer. As their names imply, these reducers convert an f/10 SCT to an f/6.3 or f/3.3 respectively. Baader Planetarium (http://www.baader-planetarium.de and http://www.alpineastro.com) makes a uniquely designed focal reducer known as the Alan Gee Telecompressor Mark II – it converts an f/10 SCT to f/5.9. Optec (http://www.optecinc.com/astronomy) makes f/5 and f/3.3 focal reducers, which they call telecompressors.

Figure 11.8. This NexStar 5i has a piggyback 35mm camera on top. Attached to the rear: an f/6.3 focal reducer; followed by an off-axis guider with illuminated reticle eyepiece; and at prime focus, a second 35mm camera mounted at the rear of the guider with a T-Ring. Photo courtesy Celestron.

SAC Imaging has created an f/7 focal reducer for the NexStar 4. This is quite an advantage over the original f/13 focal ratio. It threads to the NexStar 4's rear port and provides a standard 1.25in opening giving the flexibility to mount many of the accessories previously discussed. It is also possible to use Celestron's C90 Large Accessory Ring (LAR; Figure 11.9) in the rear port of a NexStar 4 to provide a standard SCT threading. This would allow you to use an SCT focal reducer. Additionally you would need an SCT visual back (Figure 11.1) or other suitable camera adapter. Note that an f/6.3 focal reducer actually reduces the existing focal length to 63% of the original, thus such a combination would produce a focal ratio of about f/8 on a NexStar 4.

Conclusion

Astrophotography can be a rewarding and challenging extension of your interest in astronomy, but you must be prepared for the discipline and time requirements astrophotography demands. Many beginning backyard observers feel the desire to jump into astrophotography very early on. I strongly recommend that you spend at least a year limited to visual observation before you delve into imaging. See as much as you can with your own eyes before you turn to the task of capturing photons with a camera.

Figure 11.9. This NexStar 4 has an LAR attached to the rear port, followed by an SCT T-Adapter, and then a 35mm camera mounted attached with a T-Ring. Also note the magnifier attached to the viewfinder of the camera. Photo courtesy Baader Planetarium.

Image Gallery

The astrophotos comprising Figures 11.10 through 11.23 and those included in Chapter 2 were taken with NexStar telescopes. In addition to these images you will find many more at the NexStar Group's Image Gallery – http://www.buyastrostuff.com/dons. Most of these folks are beginners – every image they produce is better than the last.

Figure 11.10. Jupiter imaged with a NexStar 8 and a digital camera. Image by Al Fugiero.

Figure 11.11. Saturn imaged with a NexStar 8 and a video astronomy camera. Image by Chris Wilkie.

Figure 11.12. Jupiter and one of its moons imaged with a NexStar 114 and a digital camera. Image by Dave Scott.

Figure 11.13. Saturn imaged with a NexStar 114 and a digital camera. Image by Dave Scott.

Figure 11.14. Close-up of the Moon imaged with a NexStar 114 and a digital camera. Image by Kevin Brett.

Figure 11.15. Saturn imaged with a NexStar 5 and a digital video camera. Image by Dr. Mario Mariani.

Figure 11.16. The Sun with close-ups of sunspot groups. Imaged with a NexStar 5, Baader AstroSolar Film, and a digital camera. Image by Dr. Mario Mariani.

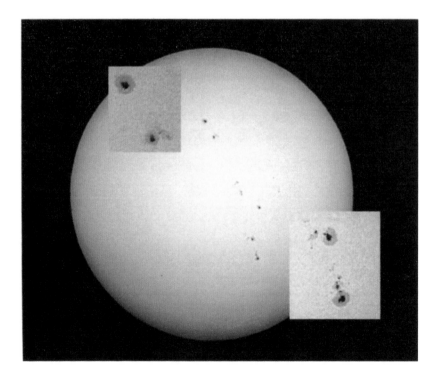

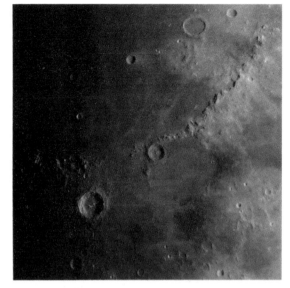

Figure 11.17. Close-up of the Moon imaged with a NexStar 5 and a digital video camera. Image by Dr. Mario Mariani.

Figure 11.18. The Crab Nebula (M1) imaged with a NexStar 11 GPS and a CCD camera. Image by Vernon Riley.

Figure 11.19. Saturn and its moon Titan imaged with a NexStar 4 and a modified webcam. Image by Stéphane Poirier.

Figure 11.20. Jupiter and its moon Europa imaged with a NexStar 5 and a digital video camera. Notice Europa's shadow on Jupiter's surface. Image by Dr. Mario Mariani.

Figure 11.21. NGC 891 imaged with a NexStar 11 GPS and a CCD camera. Image by Vernon Riley.

Figure 11.22. The Hercules Cluster (M13) imaged with a NexStar 11 GPS and a CCD camera. Image by Vernon Riley.

Figure 11.23. The Dumbbell Nebula (M27) imaged with a NexStar 11 GPS and a CCD camera. Image by Vernon Riley.

Chapter 12

Maintenance, Care, and Cleaning

As is true with any machine or instrument, a little preventive maintenance will keep your NexStar in top shape for many years to come. Depending upon your mechanical inclination, you may elect to follow some of the procedures in this chapter, or perhaps you will ask your dealer to occasionally overhaul your telescope.

Cleaning

Optical Surfaces

When it comes to cleaning optical surfaces, clean as *infrequently* as possible. Dust visible on the optics in the light of day will hardly be detectable when the scope is in actual use. A full cleaning of optical surfaces should be performed not more than once every year. If your optics don't seem very dirty after a year, don't clean them! Cleaning almost always causes some permanent damage to your optics and most foreign matter causes no harm.

When it does come time to clean keep two things in mind at all times: dust scratches and be gentle.

I do recommend cleaning as soon as possible if any liquids (other than dew) wind up on your optics. Other than that, a light dusting with a large blower bulb and soft lens brush, both available from most telescope and camera dealers, is the only regular cleaning your optics will need. I strongly recommend you do **not** use canned air. Some makes contain propellants that will adhere to your optics. On humid days, all canned air will cause very harsh condensation to form on optics.

Eyepieces are the exception to the cleaning rule. Fingerprints, high humidity due to moisture from your eyes and breathing, eye gunk, kids asking "Do you look right here?" while pointing with their chocolate-covered fingers – eyepieces seem to be magnets for contaminants. While you still need to clean as infrequently as possible, eyepieces will require more frequent cleaning than other optical surfaces.

To protect your optics, always use the dust covers and caps. Not only does this help keep clean the optical surfaces you can see, it also keeps dust out of the inside of the telescope tube – an area you will seldom want to clean.

You will need a large blower bulb, a camelhair brush, lint-free tissues, and cleaning fluid. The brush is used to very gently dislodge any dust that won't leave the surface with air from the bulb. Special lens cleaning tissues are available from most astronomy and photography dealers, and many experienced amateur astronomers use white Kleenex® tissues. Be sure not to use tissues with lotions or fragrance, only plain white tissues. And other brands do not seem to hold up as well and tend to leave lint on the surface. In no case should you use dry tissues of any type to rub your optics; scratches will result.

Cleaning fluids for glass surfaces can be purchased, but many tend to leave a film. Alpine Astronomical (http://www.alpineastro.com) stocks and supplies cleaning fluids that work quite well, or you can make your own. Mix equal amounts of distilled water and isopropyl alcohol (it is important that it is 91% alcohol or higher) and add two or three **drops** of liquid dishwashing soap – nothing fancy here, just plain dishwashing soap. A good idea for using and storing this fluid is a new spray bottle – it is best not to reuse a household cleaner spray bottle due to residue – it will be much easier to work with in this fashion.

Objective Lenses and Corrector Plates
The front glass surfaces on refractor, SCT, and Maksutov telescopes are treated with a fairly tough anti-reflective coatings. Nonetheless, repeated cleanings create microscopic scratches in these coatings that rob your telescope of some of its contrast. Use a blower bulb regularly to keep the surface clear of stray dust. A good routine to follow is to dust the optics after each use of the telescope. Dust left on the surface will become stuck to it the next time your lens or corrector is covered with dew. In fact, dew is your chief enemy in keeping these surfaces clean. Adequate dew prevention as discussed in Chapter 8 will allow you to go a long time between cleanings.

To clean your optics, follow these steps:

- Start by only cleaning on a bright, sunny day in a room with lots of light from the outside. Set up your telescope and point the front a little up from level. Set up a chair and a small table and position the height of the scope so that you can clearly see any dust on the lens or corrector when you are seated.
- Take the large blower bulb and remove all of the dust possible. For the few remaining bits, use the camelhair brush to *very gently* brush them loose. Follow with the blower bulb. **Remember:** Dust scratches.
- Next, get two tissues, fold them into quarters (this results in a 4-inch/10cm square) and set one aside. Spray (or otherwise wet) the cleaning fluid on the other tissue until about half of it is damp. Do not use so much fluid that it is dripping, just damp. Working from the center of the lens or corrector plate, **drag** the tissue to the edge. Follow immediately by dragging the dry tissue across the same area to absorb the fluid. Throw those two tissues away and repeat with another pair, working your way around the lens or corrector. If you notice any stubborn dirt, do not rub, instead, repeat with a damp and dry tissue until clean. Remember: **Be gentle**, do not apply pressure.
- After you have completed the entire surface, gently breathe on the surface and look for streaks. If there are any, repeat with a damp and dry tissue across that area. Light streaking will not generally be noticeable during actual use; do not be overly concerned with perfection.

Mirrors
The mirrors of SCT and Maksutov telescopes are quite well protected in their closed optical tubes; it will be many years before any cleaning is required. By that time, they may actually need to be recoated, a task requiring expert attention. The disassembly required to clean the mirrors will not be discussed here, but I would refer you to the excel-

lent book *Choosing and Using a Schmidt Cassegrain Telescope* by Rod Mollise for additional details.

The mirrors in a Newtonian telescope are open to the outside and will occasionally require cleaning. Again, removing a little dust may require only forced air from a blower bulb. The mirrors must be removed from the optical tube and from their supports prior to cleaning. The surface of a mirror is very delicate and easy to scratch; be careful not to touch it. Start by removing the three screws on the sides of the primary mirror cell, the black metal ring on the rear of the scope. Hold the mirror cell as you remove the screws to prevent the mirror from falling. Remove the screws on the clips holding the mirror in the cell and set the mirror aside, shiny side up. Remove the secondary support by loosening the hex screws slightly and completely removing the large center screw. Be sure to hold the secondary support to prevent it from falling when the screw comes loose. Set the secondary mirror aside.

For mirrors, you will need pure cotton balls (synthetic materials will likely scratch the surface) if there are any stubborn spots. You will also need a sink to work in, a mild dishwashing soap, and a couple of towels. Before starting, remove any rings from your fingers to avoid accidental scratches. To clean the primary mirror, follow these steps:

- Take the large blower bulb and remove all of the dust possible. Do **not** use a brush on stubborn particles; a mirror is much more delicate than a coated lens.
- Line a sink with a towel and run warm water over the surface of the mirror to remove additional dust.
- Fill the sink with about 2 inches (5 cm) of warm water. Add 2 or 3 **drops** of mild dishwashing soap. Mix the soap into the water and then set the mirror, shiny side up, in the bottom of the sink. Allow it to soak for a few minutes.
- Lift the mirror and look for remaining dirt or spots. To attempt to remove them, submerge the mirror again and gently drag a cotton ball across the surface. Do not apply any pressure at all; let the weight of the cotton do the work. Work only in a straight line; do not make a circular motion. **Remember:** Be gentle.
- After you have finished, drain the sink and run warm water over the mirror. Tilt the mirror and allow all the water to drain. If any water drops remain, dip the corner of a tissue in the drop to absorb it. Set the mirror at an angle on a large towel and allow it to completely dry before reassembling the scope.

To clean the secondary mirror, follow the directions above for lenses, being extra careful not to apply any pressure with the tissues during cleaning. Note that the secondary often does not attract as much dirt as the primary due to their orientations.

Eyepieces Eyepieces can be quite difficult to clean, especially those of shorter focal lengths with their extremely small lenses. Additionally, it is sometimes necessary to clean the bottom lens on the inside of the barrel – an especially difficult location to reach. The general procedures outlined for lenses and corrector plates are also useful for eyepieces. As usual, start with a good dusting using a bulb blower and camelhair brush. Extra care must be taken not to use too much cleaning fluid as it might leak down into the lens assembly. If this occurs, do **not** attempt to disassemble the eyepiece; you are certain to do more harm than good.

An alternative is to purchase a product known as LensPen Mini Pro. This tool has a non-abrasive cleaning tip on one end and a camelhair brush on the other. The tip is small enough for all but the smallest eyepiece lenses. Since the LensPen's cleaning tip picks up dust that may scratch lens coatings, be sure to dust well prior to use.

Diagonals As discussed in Chapter 8, diagonals come in two types: glass prism and mirror. Prism diagonals use a triangular-shaped glass prism with a side facing the eyepiece holder and a side facing the front barrel. As always, the only frequent cleaning you should perform is removing dust with a bulb blower and a camelhair brush. When more serious cleaning is necessary, I recommend using a LensPen or removing the prism and cleaning it like an objective lens.

Standard mirror diagonals are easily scratched and apart from dusting should be cleaned only rarely. The safest method is to remove the mirror and follow the mirror cleaning procedure above. Dielectric mirrors (most manufacturer's call these "enhanced") are immune to scratching during normal cleaning. Remove dust with the blower and brush and either clean with the lens procedure or use a LensPen.

Exterior Surfaces

Any cloth can be used to remove dust from the exterior of your telescope. I prefer to use a furniture polishing cloth (dry). For any stubborn dirt, use a *slightly* damp cloth. To produce a nice, clean sheen on all those stylish curved surfaces, use a plastic conditioning wipe like Armor All®. Another useful product from the car care center or boat shop is car or fiberglass wax. Applied to optical tubes, wax provides a great shine and protects the tube as well. Use care not to get wax on plastic surfaces; some waxes leave a white residue that is difficult to remove, resulting in a scope that looks worse than when you started!

Mechanical Care and Adjustments

The mechanical design of the NexStar 60/80/114/4 and the NexStar 5/5i/8/8i telescopes is actually quite simple. If you are mechanically inclined you will find it easy to completely disassemble and reassemble any of them. There are only a few mechanical adjustments on these models and they are forgiving of slight maladjustment. These models also have few internal wires, further simplifying maintenance.

The NexStar 8/11 GPS telescopes are a bit more complicated. The much more precise gears and bearings require more exacting adjustment. There are also many more wires routed throughout the base and fork arms. Some internal maintenance is simple and highly recommended, but for items beyond those mentioned here, I would recommend anyone not familiar with precise mechanical adjustment to return the scope to their dealer or Celestron if mechanical troubles arise.

Caution: The very real possibility exists during any of the procedures described here that you may damage something. Also, some of these procedures may void your warranty and you should consider returning your telescope for warranty work rather than attempting adjustments on your own. After the warranty has expired you should consider the possibility that you will cause costly damage when working on your scope. While I consider any of these procedures reasonably safe, not everyone should approach a thousand-dollar piece of equipment with a screwdriver in their hand.

That said, when working on your telescope, work in a well-lit location and start only when you have the time to work slowly and finish the job relatively uninterrupted. I strongly recommend that you NEVER have the power connected to your scope with the covers off. This is not really a safety measure for you – the electrical power inside is not dangerous – but rather one for your scope. If a screwdriver shorts two pins on a circuit board, damage is likely to result.

I have organized the various tips and procedures by related models.

NexStar 8/11 GPS

As noted in the introduction to this section, I do not recommend much by way of self-service maintenance on the mechanical parts of the NexStar 8/11 GPS. It can certainly be done, but many of the adjustments for the bearings and gears require a delicate touch that comes second nature to a mechanic but eludes others. If you are a mechanic, you will not need my assistance in any case.

For the rest of us, there is one task that must be performed and requires no adjustments, just reasonable care. The upper base rotates on three roller bearings consisting of steel wheels riding on a track in the lower base (Figure 12.1). Both the wheels and the track should be smooth and clean. When they accumulate contaminants the telescope turns with a rough "rumble" that you can feel when you rotate the scope with the clutch disengaged. This also inhibits smooth tracking and manual slewing with the hand control.

Cleaning the wheels and track is not difficult. Release the altitude clutch, move the optical tube perpendicular to the fork arms, and reengage the clutch. Using a 3.5mm or $\frac{9}{64}$-inch hex wrench, completely loosen the five screws securing the plastic base cover – the cover with the Aux, Auto Guide, and PC ports. It is not necessary to remove the screws as we will only be lifting the cover, not completely removing it. If you do remove the screws, note that the one in the back is shorter than the rest.

Carefully lift the base cover. Wires run from the cover to the base and only allow the cover to be raised about 2 inches (5 cm). Put a rolled-up sock in both gaps between the cover and each fork arm to hold it in place. Dampen a lint-free cloth (not tissue!) with alcohol. Do not make the cloth so wet that you can squeeze liquid out – the excess alcohol could rinse away the grease on the bearings.

Start by cleaning the wheels. Release the azimuth clutch. Hold the damp cloth to one of the wheels and rotate the base. Be careful to prevent the cloth from catching between the

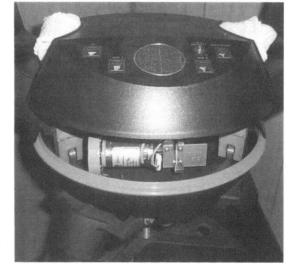

Figure 12.1. Base cover lifted to reveal roller bearings.

wheels and the base. After you clean the wheels, clean the track. Find the largest opening leading to the track and clean the track while rotating the base to gain access to the entire path.

This is the only regular maintenance procedure I recommend for the NexStar 8/11 GPS. Any time you have the base open, check all screws to insure they are tight. Also check that all accessible cables are firmly seated to the circuit boards.

After several years – the exact duration will depend largely on the amount of use your scope gets – it will become necessary to clean and regrease the bearings and gears. For most folks this will require a trip to your local scope dealer or Celestron. For the more adventuresome, take good notes as you work; maybe you can author "The Complete Maintenance Guide for the NexStar GPS"!

NexStar 5/5i/8/8i

As mentioned earlier, complete disassembly of the single-arm NexStar models is a much easier task. There are two sections that potentially need attention: the altitude axis and the azimuth axis. We will discuss disassembly, lubrication, and reassembly/adjustment. Adjustments to the altitude axis are the most common requirement as it carries the weight of the optical tube directly. You will not likely find it necessary to disassemble and lubricate your telescope for several years. I would recommend lubrication once every two or three years, unless your scope starts to exhibit problems preventing smooth motion.

Disassembly of the Altitude Axis To disassemble the altitude axis, disconnect the hand control and set it aside. Use a hex wrench to remove the three screws holding the hand control hook. It is best to loosen the screws slightly and then hold the entire telescope on its side with the fork arm down while removing the screws. This helps to prevent the screws from falling inside the fork arm. If they do, you **must** retrieve them as they will bind the gears in the base. After removing the hook, you will see a nut attached to the altitude axis as shown in Figure 12.2. Using a marker, mark the current location on the nut and the fork arm beneath it. This will allow you to tighten the nut to the same tension when you reassemble.

While holding the optical tube, remove this nut and the washers and bearings beneath. Set them on a clean cloth in the order they are removed to insure correct reassembly. It is best to make a quick sketch of the order of assembly for these parts. Slide the optical tube and altitude axis from the fork arm.

A large gear is found between the optical tube and the fork arm. It will either remain in the fork arm or stay with the optical tube. In either case, remove it as well and set it on a clean cloth. Note that the gear has three nylon disks, each in its own recess in the gear. These disks support the gear and ride in a track found in the fork arm. Occasionally the disks stay in the track; if this happens, remove them from the track and place them in the recesses in the gear.

Between the gear and the bracket attached to the optical tube you will find a fiber disk. This disk provides the slip that allows the optical tube to be moved up and down by hand while at the same time providing sufficient grip to hold the tube during normal use. This disk and the surfaces it mates to must be kept free of grease.

Disassembly of the Azimuth Axis To disassemble the azimuth axis, start by removing the battery cover. Next remove the four screws (two on either end of the

Figure 12.2. NexStar 5/5i/8/8i altitude nut with hand control hook removed. Photo courtesy Ray Cooper.

battery holder) securing the base cover to the base. When removing the base cover you will need to disconnect the cable connecting the battery holder to the base. It is keyed to insure correct polarity when you reassemble.

Under the cover you will find one (NexStar 5/8) or two (NexStar 5i/8i) circuit boards as shown in Figure 12.3. Remove these, making a careful sketch of all cable connections. Be sure to indicate wire colors on your sketch.

Under the circuit board(s) you will find a large nut attached to the azimuth axis. While holding the fork arm, remove this nut and the washers and bearings beneath. Set them on a clean cloth in the order they are removed to insure correct reassembly. It is best to make a quick sketch of the order of assembly for these parts. Lift the fork arm and upper base assembly and slide it from the azimuth axis. Set the assembly aside on a clean surface.

A large gear is found between the upper and lower base. Remove it and set it on a clean cloth. Note that the gear has three nylon disks, each in its own recess in the gear. These disks support the gear and ride in a track found in the base. Occasionally the disks stay in the track; if this happens, remove them from the track and place them in the recesses in the gear.

You may have noticed I have made no mention of a fiber disk in the azimuth assembly - there is none. The scope cannot be moved by hand in azimuth.

Cleaning and Lubrication
Start by cleaning away old grease. For removable parts that are **all** metal (the large gears, bearings, and washers) you may use an old

Figure 12.3. NexStar 5/5i/8/8i with the base cover and inside arm cover removed. Photo courtesy Mario Mariani.

toothbrush and kerosene in a small plastic basin while working in a well-ventilated area. (Please **do not use gasoline** – it is extremely flammable and the fumes are explosive!) To remove the slight residue left by the kerosene and grease, spray-on automotive break cleaner can be used. Break cleaner will deform plastic – use care! Do not immerse the fiber disk in any liquids, although you can use a clean cloth dampened *lightly* with alcohol to degrease the surfaces of the disk. Use alcohol to completely degrease the parts that mate to the fiber disk as well. For other parts, use cotton swabs, lint-free cloths, and alcohol.

Lubricate the bearings with a small but evenly applied coating of high-quality lithium or synthetic grease. Also apply grease in the recesses for the nylon disks in the large gears. A thin coating of grease is also needed in the tracks that the disks ride in. Finally, apply a small amount of grease to the teeth of the large gears.

Reassembly and Adjustment of the Azimuth Axis
Referring to your sketch, reassemble all parts in the correct order. Tighten the nut until the fork arm no longer rocks left and right. This adjustment is not as sensitive as the altitude axis, but if the nut is too tight, the scope will not slew left and right smoothly and might not slew at all at lower rates. If it is too loose, the fork arm will rock left and right. Reinstall the circuit board(s) and base cover and test the azimuth movement. Unfortunately a little trial and error may be required before you are satisfied with the results.

Reassembly and Adjustment of the Altitude Axis
While you have it apart, insure the screws holding the optical tube to the mounting bracket are tight. Occasionally these screws come loose and result in some play in the altitude axis. Also check to be sure that the three screws securing the motor (directly below the altitude axis nut in Figure 12.2) are tight.

Referring to your sketch, reassemble all parts in the correct order. Tighten the nut back to the original location. Push down on the front of the scope. If it moves too easily (some experience will be called for), tighten the nut a bit and try again. If you tighten it too much, the motor will be under excessive strain, up/down motion will not be smooth, and the scope will use too much battery power. If it takes significant effort to push the front of the scope up or down, loosen the nut a small amount. When you think you have it right, connect power and set the slew rate to 1. Check to insure the scope will move at this slow rate. Put in an eyepiece and focus on a distant object. Check that motion is smooth while observing through the eyepiece. If not, loosen the nut a very small amount and test again.

Adjustment of Motor Position
Adjusting the position of the motors requires a bit more mechanical skill than the previous adjustments. If you are not familiar with gear sets, proceed with caution and realize that adjusting the motor position will yield only a small improvement, if any, in the operation of your telescope.

The position of the altitude and azimuth motors affects the gear mesh between the final gears on the motor and the large gears attached to the axes. If the gear mesh is too loose, you will have excess play in the affected axis. If it is too tight, movement will be rough or may actually jam if the large drive gear is not perfectly circular (very common). If you notice excess play or the scope will not move through its complete range and you have ruled out other factors discussed in the axis adjustment sections, you may be able to improve the situation by adjusting the motor position for that axis. Please note that there will always be some play; if these gears are too closely meshed they will not move smoothly.

The screws holding the altitude motor in place are found behind the hand control hook. As seen in Figure 12.2, there are three screws for the altitude motor. Loosen the screws slightly, adjust the distance between the gears, and then retighten. Test the amount of free play and adjust again if necessary. After you are satisfied, power up the scope and use the up and down arrow buttons to rotate the altitude axis a complete 360°. Observe the gears and insure they do not bind. If they do, you need to allow more distance between the gears.

The azimuth motor is mounted at the base of the fork arm. This mount does not have any appreciable adjustment. If you feel there is excess free play in this axis you may elect to use a small file or hobby grinder to elongate the mounting holes to provide adjustment. You will need to disassemble the azimuth axis as described above and you may also find it easier to perform this task by removing the entire fork arm. Keep in mind that the metal filings will need to be completely cleaned from the base as otherwise they will definitely adhere to the grease lubricating the scope and cause undue wear on these parts. This by itself may be a reason for declining to attempt this procedure.

After reassembly, adjustment of the azimuth gear mesh proceeds as described for the altitude axis.

NexStar 60/80/114

Disassembly of the NexStar 60/80/114 models is easiest of all the NexStars. There are two sections that potentially need attention: the altitude axis and the azimuth axis. We will discuss disassembly, lubrication, and reassembly/adjustment. Adjustments to the altitude axis are the most common requirement as it carries the weight of the optical tube directly. You will not likely find it necessary to disassemble and lubricate your telescope for several years. I would recommend lubrication once every two or three years, unless your scope starts to exhibit problems preventing smooth motion.

Disassembly of the Altitude Axis

Start by removing the optical tube from the tube ring. Inside the tube ring you will find a large nut securing the ring to the altitude axis, as shown in Figure 12.4. While holding the tube ring, remove this nut and the washers beneath. Set them on a clean cloth in the order they are removed to insure correct reassembly. It is best to make a quick sketch of the order of assembly for these parts. Slide the tube ring from the fork arm.

A large gear is found between the tube ring and the fork arm. It will either remain in the fork arm or come off with the tube ring. In either case, remove it as well and set it on a clean cloth. Note that between the gear and the fork arm there are three nylon disks, each in its own recess in the fork arm. These disks support the gear and slide along the surface of the gear. Occasionally the disks stick to the gear; if this happens, place them in the recesses in the fork arm.

Between the gear and the tube ring you will find a fiber disk. This disk provides the slip that allows the optical tube to be moved up and down by hand while at the same time providing sufficient grip to hold the tube during normal use. This disk and the surfaces it mates to must be kept free of grease.

Disassembly of the Azimuth Axis

Start by removing the optical tube from the tube ring. Next, remove the mount from the tripod. On the bottom of the base you will find a metal bar held by two screws. Remove the screws and bar to access the large nut underneath. Remove this nut and the washers beneath. Set them on a clean cloth in the

Figure 12.4. NexStar 114 with optical tube and base cover removed.

order they are removed to insure correct reassembly. It is best to make a quick sketch of the order of assembly for these parts. Slide the base away from the rest of the mount.

A large gear is found between the base and the mount. It will either remain in the mount or come off with the base. In either case, remove it as well and set it on a clean cloth. You will also find three nylon disks in recesses around the bottom edge of the mount. These disks support the mount and slide along the top surface of the base. Occasionally the disks stick to the base; if this happens, place them back in the recesses in the mount.

Between the gear and the base you will find a fiber disk. This disk provides the slip that allows the mount to be moved left and right by hand while at the same time providing sufficient grip to drive the mount during normal use. This disk and the surfaces it mates to must be kept free of grease.

Cleaning and Lubrication

Start by cleaning away old grease. For removable parts that are **all** metal (the large gears) you may use an old toothbrush and kerosene in a small plastic basin while working in a well-ventilated area. (Please **do not use gasoline** – it is extremely flammable and the fumes are explosive!) To remove the slight residue left by the kerosene and grease, spray-on automotive break cleaner can be used. Break cleaner will deform plastic – use care! Do not immerse the fiber disks in any liquids, although you can use a clean cloth dampened *lightly* with alcohol to degrease the surfaces of the disk. Use alcohol to completely degrease the parts that mate to the fiber disks as well. For other parts, use cotton swabs, lint-free cloths, and alcohol.

Apply high-quality lithium or synthetic grease in the recesses for the nylon disks. A thin coating of grease is also needed on the surfaces that the disks slide on. Also, apply a small amount of grease to the teeth of the large gears.

Reassembly and Adjustment of the Azimuth Axis

Referring to your sketch, reassemble all parts in the correct order. Tighten the nut just to the point that the base no longer rocks back and forth. This adjustment is not as sensitive as the altitude axis, but if the nut is too tight, the scope will not slew left and right smoothly and might not slew at all at lower rates. If it is too loose, the fork arm will rock left and right. Power up the scope and test the azimuth motion. A little trial and error may be required before you are satisfied with the results.

If you cannot seem to get a good adjustment, there is a second nut on the top end of the azimuth axis. It is accessed under the plastic cover on top of the base. Along the bottom edge of this cover, on both sides near the fork arm, you will find a small opening that allows you to pry the cover up and remove it. It is attached to the base with two-sided foam tape. Under this cover you will find a nut as shown in Figure 12.4.

This nut and the one underneath the base work together to provide stability and smooth movement of the azimuth axis. The nut underneath the base adjusts the ability of the azimuth axis to slip left and right. This is only important to protect the motor and this adjustment must be tight enough to drive the azimuth axis securely. The nut found under the base cover adjusts the tension of the axis against the nylon disks. Generally, the nut underneath the base can be adjusted as tight as you like. The nut under the base cover (the nut on top) is adjusted to allow smooth movement without the fork arm rocking back and forth. It takes a little trial and error to get the best setting.

Reassembly and Adjustment of the Altitude Axis

Referring to your sketch, reassemble all parts in the correct order. Tighten the nut until the altitude axis does not rock back and forth. Push down on the front of the scope. If it moves too easily

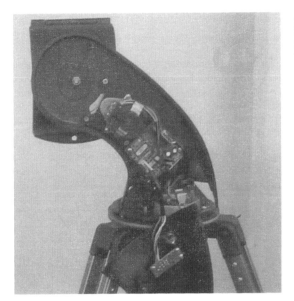

Figure 12.5. NexStar 114 with fork arm cover removed.

(some experience will be called for), tighten the nut a bit and try again. If you tighten it too much, the motor will be under excessive strain, up/down motion will not be smooth, and the scope will use too much battery power. If it takes significant effort to push the front of the scope up or down, loosen the nut a small amount. When you think you have it right, connect power and set the slew rate to 1. Check to insure the scope will move at this slow rate. Put in an eyepiece and focus on a distant object. Check that motion is smooth while observing through the eyepiece. If not, loosen the nut a very small amount and test again.

If you cannot seem to get a good adjustment, there is a second nut on the other side of the fork arm. It is found under the plastic cover on the outside of the fork arm, as shown in Figure 12.5. Remove the two screws holding the plastic cover to the outside of the fork arm. There is a cable connecting the power and hand control jacks to the circuit board found in the fork arm. You may allow the cover to hang by this cable (be careful not to damage the cable while you work) or make a sketch showing the connection of this cable to the circuit board. On the sketch be sure to note the colors of the wires.

This nut and the one inside the tube ring work together to provide stability and smooth movement of the altitude axis. The nut inside the tube ring adjusts the ability of the altitude axis to slip up and down. This is important to protect the motor, but this adjustment must be tight enough to support the weight of the optical tube. The nut found under the fork arm cover adjusts the tension of the axis against the nylon disks. Generally, the nut in the tube ring can be adjusted as tight as you like while still allowing a good push on the front of the optical tube to cause it to slip. The nut under the fork arm cover is adjusted to allow smooth movement without the optical tube rocking up and down. Unfortunately, it takes a little trial and error to get the best setting.

Adjustment of Motor Position Adjusting the position of the motors requires a bit more mechanical skill than the previous adjustments. If you are not familiar with gear sets, proceed with caution and realize that adjusting the motor position will yield only a small improvement, if any, in the operation of your telescope.

The position of the altitude and azimuth motors affects the gear mesh between the final gear on the motors and the large gears attached to the axes. If the gear mesh is too loose, you will have excess play in the affected axis. If it is too tight, movement will be rough or may actually jam if the large drive gear is not perfectly circular (very common). If you notice excess play or the scope will not move through its complete range and you have ruled out other factors discussed in the axis adjustment sections, you may be able to improve the situation by adjusting the motor position for that axis. Please note that there will always be some play; if these gears are too closely meshed they will not move smoothly.

The altitude motor is mounted in the fork arm on a metal plate. It is accessible after removing the plastic cover on the outside of the fork arm. The metal plate is secured to the fork arm with four hex screws. There is a wide range of adjustment available when these four screws are loosened. To see the spacing between the gears you will need to remove the tube ring as discussed above in the section on disassembly of the altitude axis. After adjusting the motor position, test the amount of free play and adjust again if necessary. After you are satisfied, attach and adjust the tube ring. Power up the scope and use the up and down arrow buttons to rotate the altitude axis a complete 360°. If the motor is straining, either the nut inside the tube ring is too tight or the gears are binding. If you determine that the problem is the gears, you will need to allow more distance between them.

The azimuth motor is mounted at the base of the fork arm on a small metal plate. It is accessible after removing the plastic cover on the outside of the fork arm. The metal plate is secured to the base of the mount with three hex screws. There is a wide range of adjustment available when these three screws are loosened. To see the spacing between the gears you will need to remove the base of the mount as discussed above in the section on disassembly of the azimuth axis. After adjusting the motor position, test the amount of free play and adjust again if necessary. After you are satisfied, attach and adjust the base. Power up the scope and use the left and right arrow buttons to rotate the azimuth axis a complete 360°. If the motor is straining, either the nut underneath the base is too tight or the gears are binding. If you determine that the problem is the gears, you will need to allow more distance between them.

NexStar 4

The NexStar 4 is the most difficult of the little NexStars to disassemble and adjust correctly. There are two sections that potentially need attention: the altitude axis and the azimuth axis. We will discuss disassembly, lubrication, and reassembly/adjustment. Adjustments to the altitude axis are the most common requirement as it carries the weight of the optical tube directly. You will not likely find it necessary to disassemble and lubricate your telescope for several years. I would recommend lubrication once every two or three years, unless your scope starts to exhibit problems preventing smooth motion.

Disassembly of the Altitude Axis First remove the hand control bracket located in the fork arm. It is best to loosen the screws slightly and then hold the entire telescope on its side with the fork arm down while removing the screws. This helps to prevent the screws from falling inside the fork arm. If they do, you **must** retrieve them, as otherwise they will bind the gears in the base. After removing the hook, you will see a nut attached to the altitude axis as shown in Figure 12.6.

Remove this nut while supporting the optical tube. Lay the nut and washers on a clean cloth in the order in which you remove them. It is critical that they are reassembled in the same order. It is best to make a quick sketch of the order of assembly for these parts. At

Figure 12.6. NexStar 4 with hand control bracket and inside arm cover removed. Photo courtesy Dan Hupp.

this point, the optical tube, a large gear, and the shaft they both ride on will come free from the fork arm. You will also find three nylon disks held in three recesses on the large gear. If any of the nylon disks is missing, you will find it in the round track inside the fork arm. When reassembling, you must position these disks in the recesses in the gear.

Disassembly of the Azimuth Axis On the bottom of the base you will find a large nut. Remove this nut and the washers beneath. Set them on a clean cloth in the order in which they are removed to insure correct reassembly. It is best to make a quick sketch of the order of assembly for these parts. Slide the base away from the rest of the mount.

A large gear is found between the base and the mount. It will either remain in the mount or come off with the base. In either case, remove it as well and set it on a clean cloth. You will also find three nylon disks at the end of three posts on the bottom of the mount. These disks support the mount and slide along the top surface of the large gear. Occasionally the disks stick to the gear; if this happens, place them back on the posts in the mount.

Between the gear and the base you will find a fiber disk. This disk provides the slip that allows the mount to be moved left and right by hand while at the same time providing sufficient grip to drive the mount during normal use. This disk and the surfaces it mates to must be kept free of grease.

Cleaning and Lubrication
Start by cleaning away old grease. For the large gear removed from the base you may use an old toothbrush and kerosene in a small plastic basin while working in a well-ventilated area. (Please **do not use gasoline** – it is extremely flammable and the fumes are explosive!) To remove the slight residue left by the kerosene and grease, spray-on automotive break cleaner can be used. Break cleaner will deform plastic – use care! Do not immerse the fiber disk in any liquids, although you can use a clean cloth dampened *lightly* with alcohol to degrease the surfaces of the disk. Use alcohol to completely degrease the parts that mate to the fiber disk as well. For other parts, use cotton swabs, lint-free cloths, and alcohol.

Apply high-quality lithium or synthetic grease in the recesses for the nylon disks. A thin coating of grease is also needed on the surfaces that the disks slide on. Also, apply a small amount of grease to the teeth of the large gears.

Reassembly and Adjustment of the Azimuth Axis
Referring to your sketch, reassemble all parts in the correct order. Tighten the nut just to the point at which the base no longer rocks back and forth. This adjustment is not as sensitive as the altitude axis, but if the nut is too tight, the scope will not slew left and right smoothly and might not slew at all at lower rates. If it is too loose, the fork arm will rock left and right. Power up the scope and test the azimuth motion. A little trial and error may be required before you are satisfied with the results.

If you cannot seem to get a good adjustment, there is a second nut on the top end of the azimuth axis. It is accessed under the battery holder on top of the base. Gently pry the battery holder from the base, using care not to break the wires. Under this cover you will find a nut as shown in Figure 12.7.

Figure 12.7. NexStar 4 with base cover removed. Photo courtesy Dan Hupp.

This nut and the one underneath the base work together to provide stability and smooth movement of the azimuth axis. The nut underneath the base adjusts the ability of the azimuth axis to slip left and right. This is only important to protect the motor and this adjustment must be tight enough to drive the azimuth axis securely. The nut found under the base cover (on the top of the axis) adjusts the tension of the axis against the nylon disks. Generally, the nut underneath the base can be adjusted as tight as you like. The nut under the base cover (on top) is adjusted to allow smooth movement without allowing the fork arm to rock back and forth. It takes a little trial and error to get the best setting.

Reassembly and Adjustment of the Altitude Axis Referring to

your sketch, reassemble all parts in the correct order. Tighten the nut until the altitude axis does not rock back and forth. Push down on the front of the scope. If it moves too easily (some experience will be called for), tighten the nut a bit and try again. If you tighten it too much, the motor will be under excessive strain, up/down motion will not be smooth, and the scope will use too much battery power. If it takes significant effort to push the front of the scope up or down, loosen the nut a small amount. When you think you have it right, connect power and set the slew rate to 1. Check to insure the scope will move at this slow rate. Put in an eyepiece and focus on a distant object. Check that motion is smooth while observing through the eyepiece. If not, loosen the nut a very small amount and test again.

If you cannot seem to get a good adjustment, there is a second nut located under the optical tube. This nut and the one on the outside of the fork arm work together to provide stability and smooth movement of the altitude axis. The nut under the optical tube adjusts the ability of the altitude axis to slip up and down. This is important to protect the motor, but this adjustment must be tight enough to support the weight of the optical tube. The nut found on the outside of the fork arm adjusts the tension of the axis against the nylon disks. Generally, the nut under the optical tube can be adjusted as tight as you like (while still allowing a good push on the front of the optical tube to cause it to slip) and the nut on the outside of the fork arm is adjusted to allow smooth movement without allowing the optical tube to rock back and forth.

It is a major task to access the nut under the optical tube and a frustrating ordeal to adjust correctly. Remove the outside nut (the one under the hand control hook) and pull the optical tube from the fork arm again. Next you must remove the gear from the optical tube. It is necessary to rotate the gear to align the three holes with the screws below. The best way to do this is to put the assembly back in the fork arm and push down a bit on the optical tube. This causes the optical tube to slip while the gear remains in place. Check to see if the holes and screws have lined up; if not, try again. Once you can access the screws, remove them and the OTA will finally come off the bracket attached to the gear.

At this point, put the gear, optical tube bracket, and altitude axis back into the fork arm. Tighten the nut shown in Figure 12.8 by a quarter turn. Pull the bracket and nut from the fork arm and reattach the optical tube. Reassemble everything and attempt to adjust the axis for smooth motion. It takes a little trial and error to get the best setting.

Caring for the Electronics

While the computers and motors in a NexStar telescope provide the technological wonder of its GoTo system, in many ways you are also at their mercy. None of the NexStar models is truly usable without the electronics functioning properly. Fortunately, modern electronics are very reliable. With minimal care you can expect the electronics in your telescope to last for many years. Here are some tips:

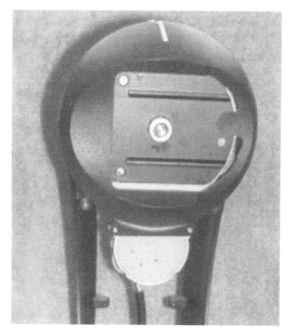

Figure 12.8. NexStar 4 with optical tube removed. Photo courtesy Dan Hupp.

- The most important measure you can take is to be certain you are using a proper power source. Several options were discussed in Chapter 8, but to review the main points, use only a 12V d.c. power source with the correct connector. It is crucial that the connector is wired for positive (+) in the center and negative (–) on the outside. Reversing polarity can yield disastrous results.

- Never disconnect or connect the hand control with the power applied. While this generally causes no problems, it can and has damaged some hand controls.

- Do not store your telescope in a humid location. High humidity will cause corrosion on electronic contacts, causing unusual misbehavior. Other metal parts will also suffer.

- Use care when connecting accessories to the "12V OUPUT" jack on the base of the NexStar 8/11 GPS. It is only designed for very low power draw. Celestron states that nothing rated higher than 750 mA should be connected to this jack and I recommend a more conservative 500 mA. Too much power and you will damage the slipring inside the base, resulting in a costly repair. This definitely rules out dew strips, but small accessories like digital/CCD cameras and focus motors (if they run off 12V) should be just fine. You will be required to make your own power cables though. Note that if you connect multiple devices, you must **add** their power ratings and insure they do not exceed 750 mA.

- Do not allow liquids to enter the hand control or telescope base. I know, you would never do that ... but that cup of hot coffee has been known to end up in places other than my stomach. If this does happen, power off immediately and pack it up for the night. When you have everything in a well-lit location, remove any covers to allow the liquid to dry more quickly. If you can, wipe up as much as possible. If you spill something into the hand control it will likely foul the button contacts. Remove the hand control case (there are six small screws on the back), using care with the wires. Pick up as much of the liquid as possible with tissues and then wait for a day for everything to dry. Then, using alcohol and cotton swabs, clean the front of the circuit board (the keypad side) and the bottom of the white rubber keys.

- If you suspect corrosion on electrical contacts, use cotton swabs and a can of contact cleaner (available at electronic parts suppliers) to clean all metal contacts. In addition to the contacts in the jacks on the base, there are cable connectors inside the base. In most cases simply slipping the connector off and on a couple of times can successfully clean these contacts. Whether this method works or you must resort to contact cleaner, it is **vital** that you reattach these connectors exactly as they were originally. Most of them have multiple pins and are easily connected with the pins offset from the intended position – be very careful.

For all procedures in this chapter, use appropriate care and **take your time**. If you run into something you didn't expect, ask for help. In addition to contacting me via my web site (http://www.NexStarSite.com), you will find the members of the NexStar and NexStarGPS Groups on Yahoo are always happy to help a NexStar owner in need.

Chapter 13
Mounting Other Optical Tubes on a NexStar

The NexStar mounts have proven to be very popular. GoTo, tracking, light weight, and a low price (particularly the GT models) make them very attractive as potential mounts for a wide variety of optical tube assemblies (OTAs). General concerns to keep in mind include:

- A longer OTA may affect your ability to point at objects high overhead.
- A longer OTA may cause balance problems and thus stress the motors or impair tracking accuracy.
- An OTA weighing much more than the original may strain the motors, shortening their life.

Each model of NexStar presents different possibilities, but do be aware that some of the following suggestions may void your warranty.

NexStar 8/11 GPS

Removing the original OTA and replacing it with another is not really an option on the NexStar 8/11 GPS. However, mounting short-focal-length refractors on the top of these scopes is very popular (Figure 13.1). The gears and motors are very sturdy and can easily take the added weight of a 3- or 4-inch wide-field refractor. The NexStar optical tube provides the high-resolution and light-gathering capabilities required for most objects while the short-tube refractor provides stunning wide-field views of open clusters and other extended objects. It is truly the best of both worlds.

Tele Vue, William Optics, and Stellarvue all make wonderful refractors suited for this purpose. Many use the popular Chinese 80mm short-tube refractor available from Orion, Celestron (the NexStar 80 is an example), and others. Suitable tube rings and mounting rails as discussed in Chapter 8 will be required.

Due to the height of the refractor's eyepiece, you will generally need to rotate the diagonal to the side or use the eyepiece directly in the focuser tube without a diagonal. Another possible option would be a 45° diagonal rather than the standard 90° model. Be aware, though, that many 45° diagonals degrade the image quality.

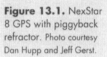

Figure 13.1. NexStar 8 GPS with piggyback refractor. Photo courtesy Dan Hupp and Jeff Gerst.

NexStar 5/5i/8/8i

For the NexStar 5/5i/8/8i mount (they are mechanically the same), the best option is "Ray's Bracket" (Figure 13.2). Available at BuyAstroStuff.com (http://www.buyastrostuff.com), Ray's Bracket is comprised of two metal rails. One rail is attached to the fork arm in place

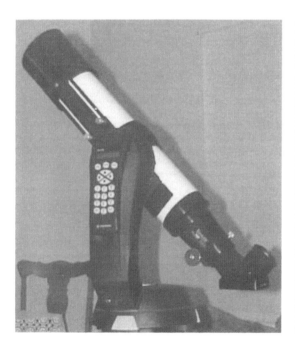

Figure 13.2. Tele Vue Genesis on NexStar 8 using Ray's Bracket. Photo courtesy Hank Williams.

Figure 13.3. NexStar 8 mounted with Ray's Bracket. Photo courtesy Frank Dilatush.

of the original OTA. The second rail slides into the first rail and is held in place with two thumbscrews for no-tools mounting of the OTA. Any suitable OTA can be attached to the sliding rail and additional rails can be purchased for multiple OTAs. In fact, if you have more than one OTA with rails attached, you can swap tubes in less than a minute. Additional photographs and the complete installation instructions are available at BuyAstroStuff.com's web site.

The motors and gears on the NexStar 5/5i/8/8i are quite robust. Well-balanced tubes up to 20 lb (9.1 kg) should not pose much trouble, although you may need to upgrade your tripod. For many optical tubes you will also need tube rings, available from most astronomy equipment dealers, to attach the tube to the rail.

Swapping optical tubes is not the most common usage of Ray's Bracket – the 5in and 8in SCT Celestron optical tubes are hard to beat. A more common reason for using Ray's Bracket is to take advantage of the ability to slide the OTA backward and forward. This can be critical to improving the balance of the scope with heavy equipment attached. For example, a large digital camera mounted at the eyepiece may cause tracking to suffer. With Ray's Bracket, you can slide the OTA forward to improve balance and tracking. Also, as mentioned in Chapter 8, a 2-inch diagonal and other equipment may not clear the base on a NexStar 8/8i when pointed towards the zenith. Ray's Bracket allows the OTA to slide forward, easily providing the needed clearance (Figure 13.3).

NexStar 4

Options are limited with the NexStar 4; there are currently no adapters or brackets on the market, although the Baader Bracket discussed in the next section can be successfully modified for the NexStar 4. If you are determined to mount another OTA on a NexStar 4, you will need to disassemble the current mounting bracket and fashion something on your own. Metalworking skills will likely be needed, although a wooden bracket could be fashioned. Be aware before you start that the motors and gearing in the NexStar 4 are much less robust than the larger NexStars. Short-tube refractors and small Maksutov telescopes weighing no more than about 10 lb (4.5 kg) should work fine, although pointing to the zenith may be restricted by contact with the base. Balancing the weight of the optical tube and accessories is crucial to achieve acceptable tracking. The following instructions can get you started, but proceed at your own risk.

To remove the NexStar 4 optical tube, you must first remove the hand control hook located in the fork arm. It is best to loosen the screws slightly and then hold the entire telescope on its side with the fork arm down while removing the screws. This helps to prevent the screws from falling inside the fork arm. If they do, you **must** retrieve them as they will bind the gears in the base. After removing the hook, you will see a nut attached to the altitude axis.

Remove this nut while supporting the optical tube. Lay the nut and washers on a clean cloth in the same order you remove them. It is critical that they are reassembled in the same order. At this point, the optical tube, a large gear, and the shaft they both ride on will come free from the fork arm. You will also find three nylon disks held in three recesses on the large gear. These, in effect, are the bearings of the altitude axis. If any of the nylon disks are missing, you will find them in the round track inside the fork arm. When reassembling, you must position these disks in the recesses in the gear.

To remove the gear from the optical tube, you must rotate the gear to align the three holes with the screws found below. The best way to do this is to put the assembly back in the fork arm and push down a bit on the optical tube. This causes the optical tube to slip while the gear remains in place. Check to see if the holes and screws have lined up; if not, try again. Once you can access the screws, remove them and the OTA will finally come off the bracket attached to the gear. At this point, you are on your own with fashioning a method of attaching an alternate optical tube.

NexStar 60/80/114

For the NexStar 60/80/114, the easiest method is to mount another OTA in the existing tube ring. (Binoculars may be mounted on PVC tubing held in the tube ring – see Figure 13.4.)

Figure 13.4.
Binoculars mounted on PVC tubing held in tube ring of a NexStar 80.
Photo courtesy Joe Shuster.

Figure 13.5. William Optics Megrez 80 mounted directly in the tube ring of a NexStar 80. Photo courtesy Don Wyman.

The inside diameter of the N114 tube ring is 140 mm, that of the N80 (Figure 13.5) tube ring is 90 mm, and that of the N60 tube ring is 63 mm. OTAs slightly larger would fit just fine. OTAs slightly smaller could be mounted by using felt. Please note that the diameter of a scope is almost never the same as its aperture, so be sure to check the actual size before ordering that OTA of your dreams.

The NexStar 60/80/114 models use the same motors as the NexStar 4 and the same recommendations for suitable optical tubes are true here as well. Short-tube refractors and small Maksutov telescopes weighing no more than about 10 lb (4.5 kg) will work fine. Balancing the weight of the optical tube and accessories is crucial or tracking will suffer. In fact, it is possible to mount a heavier optical tube, if the balance is nearly perfect.

For a wider range of possibilities there is the Baader Bracket (Figures 13.6 and 13.7). Baader Planetarium has designed a wonderful bracket available in Europe from Baader (http://www.baader-planetarium.de) or in North America at Alpine Astronomical (http://www.alpineastro.com). The bracket has two major parts – a mount with hand-screw attached to the NexStar fork arm and a dovetail bar attached to the optical tube. The dovetail is the same dimensions as those used on many German equatorial mounts, so an optical tube in your closet might be compatible with no modifications.

The dovetail bar slides into the mount and the hand-screw is tightened to secure it. The bar allows a modest amount of adjustment fore and aft to attain balance. Extra dovetail bars can be purchased for mounting additional optical tubes allowing quick and easy swapping, even after an alignment is performed. For many optical tubes you will need tube rings, available from most astronomy equipment dealers, to attach the tube to the dovetail bar. Additional photographs as well as the installation instructions are available on the Baader Planetarium and Alpine Astronomical web sites.

If you mount a refractor or a compound scope (SCT or Maksutov) on a NexStar 114 mount, go to Model Select on the Menu button and change the setting to NexStar 4. This will provide the correct gear ratio (the NexStar 4 and 114 are the same in this respect) while reversing the up/down arrow buttons at lower slew rates. This reversal makes the arrow buttons behave more naturally in the right-side-up, mirrored view of refractors and compound telescopes.

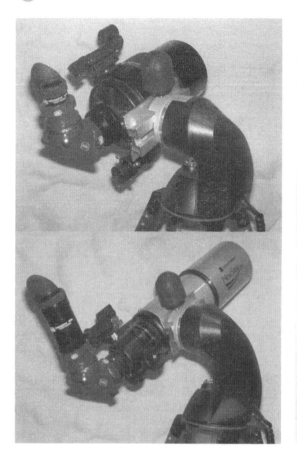

Figure 13.6.
Celestron C90 and
original NexStar 80
using Baader Bracket.
Photos courtesy Matthias
Bopp.

Figure 13.7. Orion
StarMax 127 on
NexStar 80 using
Baader Bracket. Photo
courtesy Jim Blackwell.

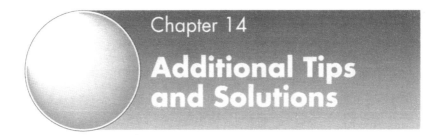

Chapter 14

Additional Tips and Solutions

Although the NexStar line of telescopes is about as easy as it gets when it comes to computerized telescopes, there is a learning curve. I hope that by now most of your questions have been answered in the previous chapters, but there are some issues yet to discuss. The NexStar line is not entirely without problems, but fortunately many of the potential difficulties have simple solutions. When all else fails, it is time to contact Celestron Technical Support at (310) 328-9560 or via email at support@celestron.com.

The tips in this chapter are divided into sections for specific NexStar models and a section for tips that apply to multiple models. Refer to Chapter 3 for clarification regarding the original/new GT hand control and the various versions of the GPS hand control/motor control.

Various Models of NexStar

Buying a Used NexStar

(All models)

Amateur astronomers generally take excellent care of their equipment. Buying a used telescope is a good way to save substantial money while still getting the scope of your dreams. Here are some things to check on a used NexStar.

- Look for any obvious damage. Even if the dents or scrapes do not seem to have damaged any critical components, the impact that caused them may have caused unseen damage.
- Check the surfaces of the mirrors/lenses for scratches or discoloration that will impair optical performance.
- Try the scope out at night, checking for pinpoint star images and similar inside- and outside-focus star patterns as discussed in Chapter 9. Keep in mind that distorted patterns can be caused by poor collimation or inadequate time to cool the optics. If you cannot check the optics at night, focus on a distant object, looking for obvious defects. You can even check defocused "star" patterns using sunlight reflected off a distant glass or metal object.

- Check the focuser motion – it should move easily with little backlash.
- Power up the scope and attempt an Auto Alignment. Insure the scope slews smoothly to the alignment stars. Accept the given location of the two alignment stars. Press the Tour button and select an object, pressing Enter to GoTo the object. Again, observe whether the movement is smooth.
- Set the rate to 2 and focus on a star or distant object. Does the scope move smoothly (although slowly) in all four directions?

If the scope checks out, consider the age and condition and compare with the cost of a factory new telescope. With the number of NexStar telescopes sold since their recent introduction, I expect the market for used models to be substantial in the next few years. Personally, I would hesitate to purchase a used telescope via the Internet or through the mail – there is often little recourse if the telescope happens to be damaged – but if you can see and test it in person, a used scope is often a great deal.

Some of the Features Described in This Book Are Not Found on Your Scope

(All models)

Check to see if I noted that the feature applies to your model of NexStar. If so, perhaps you have an older version of hand control or motor control firmware. Refer to Chapter 3 for information on models and versions. If it is not possible to upgrade to a newer version, some of the tips in this chapter will help to overcome problems with older versions.

No Power When Running on Internal Batteries

(NexStar 5/5i/8/8i)

The battery compartment on the NexStar 5/5i/8/8i holds the batteries very tightly, sometimes resulting in difficulties in making a good connection. After inserting new batteries, if the telescope will not power up, try reseating the batteries. The best solution is to remove the batteries altogether and power your scope with an a.c. adapter or external power source as described in Chapter 8.

Power-Related Problems

(All models)

Low or marginal power causes any number of problems with NexStar telescopes. Hand control lockup, GoTos that either don't reach their target or never stop looking, "No Response 16/17" errors, and various other misbehaviors of the scope are all common with faulty power. Celestron recommends an a.c. adapter with a minimum current rating of

1500 mA, but due to variations in quality I recommend a minimum of 2000 mA. Even better is a rechargeable battery pack as described in Chapter 8. Such batteries are capable of much more than 2000 mA and the added portability is appreciated even when you are close to a power outlet.

Also, momentary loss of power is a frequent culprit. The most common cause for momentary loss of power is a loose power connector. If you have problems with the power cord pulling loose when the scope slews, attach it to the fork arm with a large rubber band or a self-adhesive hook. Sometimes even this may not be enough. Bumping the power connector may continue to cause problems.

Many power cords seem to provide a good fit, but the inside hole of the connector is too large for the center post in the scope's power connector. One possible solution is to use a jeweler's screwdriver or a razor knife to **gently** spread the center post of the power socket on the base of the telescope. Be very careful not to twist the post as you perform this delicate operation – you could break the connector inside the scope, necessitating replacement of the entire power socket. A better solution is to replace the connector on the power cord. The correct connector for NexStar scopes has an outside diameter for 5.5 mm and an inside diameter of 2.1 mm. While a loose-fitting connector might seem to work, hard-to-diagnose problems often occur.

Alternate Locations for The Hand Control

(All models)

Although it was nice of Celestron to give us a place to store the hand control in the fork arm on the N4 and larger NexStars, I doubt many of us find the location convenient for operating the scope. For all models of NexStar scopes, strategically placed Velcro® will make life with your scope more comfortable.

On my NexStar 11 GPS I have attached the hand control to the base using $\frac{3}{4}$-inch self-adhesive Velcro®. Use the hook part of the Velcro® to cover the NexStar emblem on the base. Put 3-inch-long strips of the fuzzy, loop part of the Velcro® on the back of the hand control, just below the middle screws.

With the hand control plugged into Aux 2 (it works in either Aux port), the cord makes a graceful loop up to the top of the hand control when it is stuck in place. In this location you can see the hand control from your position at the eyepiece (Figure 14.1). Plus, you can put patches of Velcro® at other locations if you like! Note that the scope won't swing back to the parked position without pulling the hand control free and storing it in the holder in the fork arm. I no longer use the hand control in the fork arm position, so I never reattach the plug-in that hard-to-reach jack.

Additionally, you might consider Ray (RJ) Rauen's Keypad Holder. Models are available for the NexStar 8/11 GPS and the NexStar 5/8. See RJ's web site – http://www.hometown. aol.com/rjrjr001 – for more details.

Transporting Your SCT NexStar

(NexStar 8/11 GPS, NexStar 5/5i/8/8i)

The heavy primary mirror of a Schmidt–Cassegrain telescope rides on a metal tube (the baffle tube) secured to the rear cell of the telescope. When transporting your SCT, it is best

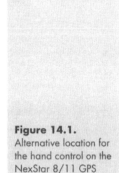

Figure 14.1.
Alternative location for
the hand control on the
NexStar 8/11 GPS

to move the mirror completely to the rear by turning the focuser knob clockwise to the end
of its travel. This minimizes the stress placed on the baffle tube by the heavy weight of the
primary mirror.

To aid the longevity of the clutches on the NexStar 8/11 GPS, it is best to release both
clutch levers after the scope is positioned for travel.

Incorrect Date and Time Displayed on GPS Models During Quick Align and Auto Align

(NexStar 8/11 GPS, NexStar 5i/8i with optional GPS module)

During a Quick Align or Auto Align the scope references the GPS module for the correct
date and time. If the GPS module can obtain a link, the date and time will be updated; if
not, the GPS module reports the date and time based on the last link, potentially off by a
few seconds. The GPS module uses a small rechargeable battery to maintain the date and
time as well as the most recent data elements downloaded from the GPS satellites. If your
telescope has not been used in several weeks or is only used infrequently for short periods
of time then this battery will loose its charge. In this case the date and time will revert to an
arbitrary state.

To recharge the battery, use an a.c. adapter and switch the power on. There is no need to
perform an alignment – in fact it is better that you don't so that the scope will not begin
tracking – but leave the power on for at least 48 hours. Then take the scope outdoors to
allow a GPS link. If the date and time are incorrect the next time you are unable to get a
GPS link (typically only if testing indoors), the battery may need replaced.

Park Position for Your Scope

(All models)

Most SCT and Maksutov owners store their scopes with the front pointed down. With the other NexStar models, other orientations may be more favorable for your uses. In either case, wouldn't it be nice if you could simply press a few buttons on your hand control and the scope would GoTo its storage position? Actually, a user-defined Land Object can be used to do just that.

The next time you are finishing up a session with your scope, use the arrow buttons to move to your desired park position. Then press the Menu button, select User Objects, and then Save Land Object. Store the current location in the first Land Object. In the future you can park the scope with GoTo Land Object.

Note that while you can release the clutches on the NexStar 8/11 GPS to move to a park position, this will loose the alignment if you intend to use the Hibernate mode. In this case, GoTo the saved Land Object, then activate the Hibernate mode.

Attaching Accessories to the Rear of Schmidt–Cassegrain Models

(NexStar 8/11 GPS, NexStar 5/5i/8/8i)

The SCT models all provide a standard SCT thread on the rear of the telescope. Many accessories are made for this threading. One accessory, the 1.25in visual back, is included with the scope. Other accessories, discussed in Chapters 8 and 11, include:

- focal reducers;
- 2in adapters – also known as 2in visual backs;
- thread-on 2in diagonals;
- SCT T-Adapters;
- off-axis guiders;
- flip mirrors;
- micro-focusers.

The NexStar 11 has a larger rear opening than the 5- and 8-inch models. An adapter that attaches to the larger 3.29in opening supplies the standard SCT threading. The only common accessory that attaches directly to the larger opening is a 2in adapter that gives a larger clear aperture than the standard SCT 2in adapter.

When the Scope Slips or Moves In Altitude

(NexStar 5i/8i, NexStar 60/80/114/4)

When using heavy eyepieces or cameras with your telescope, the scope might slip up or down in altitude. If this occurs you must power off and realign your scope. If the scope does not otherwise move easily by hand when a standard eyepiece is used, the best way to fix the problem is to add weight to the other end of the scope to achieve better balance. If you use Ray's Bracket or the Baader Bracket you can shift the optical tube forward or backward to balance the scope. With the NexStar 60 and 114 you may find there is enough latitude with the tube ring to move the optical tube to a balance point.

If the scope moves easily by hand, the altitude axis must be adjusted. Refer to Chapter 12 for the adjustment procedure.

When the Scope Won't Move in Altitude or Azimuth when Pressing Hand Control Buttons

(NexStar 5i/8i, NexStar 60/80/114/4)

Occasionally a new telescope will not move freely when using the arrow buttons on the hand control, particularly when using lower slew rates. The best solution is to return the scope to the dealer for adjustment or a replacement. The problem is generally caused by a bad motor, the axis being adjusted too tight, or the motor having been positioned too close to the large gear that drives the axis. Further information about adjusting the axes and motors can be found in Chapter 12.

Scope Rocks Left and Right

(NexStar 5i/8i, NexStar 60/80/114/4)

If the fork arm or the optical tube rock left and right, either the altitude or the azimuth axis is too loose. Refer to Chapter 12 for details on adjusting the axes.

Can the Scope be Moved by Hand?

(All models)

The NexStar 60/80/114/4 can be moved up and down in altitude and left and right in azimuth with sufficient force. This is not a practical means to use the scope without power, but rather is designed to protect the motors. If you move the scope by hand after alignment, the alignment is lost and you must power off to start again.

The NexStar 5/5i/8/8i can be moved up and down in altitude, but cannot be moved left and right in azimuth. Again this is a feature designed to protect the altitude motor, not to permit use without power. If you move the scope by hand after alignment, the alignment is lost and you must power off to start again.

The NexStar 8/11 GPS has clutches for both the altitude and azimuth axis. When the clutches are released the telescope moves freely in both axes. If you release either clutch after alignment, the alignment is lost and you must power off to start again. If you sufficiently balance the scope you might find it possible to use the telescope with the clutches released – manually pointing at objects and nudging the back of the scope to track the motion of the sky. I suspect you would rather keep a reliable power source handy.

The Image in the Eyepiece Shifts when Focusing SCT Models

(NexStar 8/11 GPS, NexStar 5/5i/8/8i)

Schmidt–Cassegrain telescopes adjust focus by sliding the large primary mirror back and forth along the baffle tube that extends into the optical tube. This results in a certain amount of lateral movement that causes the image in the eyepiece to shift. Celestron's focuser mechanism is an excellent design that should result in a minimum of image shift, however, when the scope is new the thin film of grease found on the baffle tube often needs to be distributed evenly. Take a few moments and run the focuser knob clockwise until you reach the end of travel, then counterclockwise to the other end. Repeat this two or three times and you should see an improvement in image shift.

If you feel the shift is still excessive you might loosen the three screws found around the focuser and re-center the knob. This might take some trial and error, but if the knob is off-center, the mirror is pulled laterally when focusing.

Another solution is to add a "zero-image shift" focuser between the SCT thread on the rear of the scope and the visual back or camera adapter. You then use the standard focus knob to reach rough focus and the zero-image-shift focuser for fine focus. Zero-image-shift focusers are available in motorized and manual models. Jim's Mobile, Inc. (JMI), Van Slyke Engineering, and Apogee Incorporated offer a variety of zero-image-shift focusers. Zero-image-shift focusers are of little benefit for visual use, but can be very effective for imaging.

A related trick is to always move to focus by turning the knob counterclockwise. This moves the mirror forward, or up against gravity and mirror shift is also in a uniform direction. If you focus with a clockwise motion, the mirror will gently settle a bit after a few moments, ruining the focus.

Improving Focuser Sensitivity

(All models)

Focusing at lower powers is quite easy due to a larger "sweet-spot". Focusing at high power can be a little frustrating due to the much greater sensitivity required. On almost any telescope you can improve focuser sensitivity by creating a larger focuser knob. The basic idea is to use a large circular object with a hole in the middle to fit over the existing focuser knob. Plastic jar lids, foam, model car tires – the possibilities are endless.

One immediate improvement you can make with your SCT NexStar or NexStar 4 is to pull the rubber focuser knob out far enough for it not to drag on the rear of the scope. Celestron's SCT focusing mechanism is remarkably smooth and many find this simple improvement sufficient to allow good focus control at higher powers.

The NexStar 60, 80, and 114 use a traditional rack-and-pinion focuser mechanism that has some adjustment. The screws on the bottom of the focuser assembly provide some adjustment on the tension of the focuser knob. If they are too tight, the focuser is difficult to move and it is easy to overshoot proper focus. If they are too loose, the focuser might move when you touch the eyepiece.

On the NexStar 80, there are two tiny holes on the top of the focuser assembly, one in front and one behind the large thumbscrew. The thumbscrew is used to lock the focuser to support heavy cameras that might cause the focus to shift. Inside the two holes there are very small hex screws. All three of these screws press down on a plastic bar that slides against the focuser tube. Loosen the thumbscrew and then adjust the two hex screws to provide smooth focuser motion that stays put with your heaviest eyepieces.

Another possibility is a focusing motor. Some are made to simply move the focuser knob for you. This type is made to be more or less universal for a specific type of scope: SCT, Newtonian, refractor. It is just a matter of finding one that will physically fit on your telescope. Most people quickly tire of these and simply perfect their focusing "touch". A second type is the motorized zero-image-shift focuser mentioned previously. For SCTs these can be quite useful for that final, critical focusing. There is the disadvantage of achieving rough focus with the standard focus knob and then moving to the control box for the motorized focuser.

An alternative for some SCT NexStars is an innovative focuser: the FeatherTouch SCT MicroFocuser from Starlight Instruments (sold at Starizona). This is a dual-knob focuser that fits on the original focuser shaft of Celestron 8in and 11in SCTs. One knob focuses normally; the other knob provides a 10:1 gear ratio reduction to provide extremely fine control. Though completely manual, the 10:1 reduction has become a bit of a hit with astrophotographers. And for visual use, this is about as good as it gets.

Objects Drift out of the Field of View While Tracking

(All models)

First, here is what you can expect. The NexStar 8/11 GPS and the NexStar 5i/8i should track objects very well, keeping them in the field of view for hours. The older NexStars are another story. The NexStar 60/80/114/4/5/8 were designed mainly for visual use only. They vary greatly in their tracking ability, even when comparing scopes of the same model. It is not uncommon for objects to drift out of the field of view in as little at 10 minutes.

To get an idea of what we should expect, the NexStar Group on Yahoo developed a "Standard Tracking Test" that you can execute without even seeing the sky. As the chief protagonist in this discussion, I have posted the test and many sets of results at my web site – http://www.NexStarSite.com – feel free to visit the site and compare your telescope with the posted results.

One factor that can affect tracking in all models is the anti-backlash setting. If anti-backlash is inaccurate, an object can drift out of the field of view right after you manually center it. For example, in the Northern Hemisphere most areas of the sky are moving to the right in the sky. If you manually center an object with a final movement to the left, anti-backlash should rewind the motor to the right after you release the left arrow button. This re-engages the gears to allow tracking to start immediately. If your anti-backlash settings are too low, the motor is not rewound far enough and tracking does not start immediately, causing the object to drift out of the eyepiece. Refer to Chapter 5 for the correct method to adjust anti-backlash.

Telescope Slews to Object, but Jumps Out of Field of View at the End of GoTo

(All models)

One common cause for this behavior is Negative anti-backlash set too high. Refer to Chapter 5 for the correct method to adjust anti-backlash. Another potential cause is an extreme balance problem. Try adding weight to the front (most commonly required) or the back of the telescope to improve the situation.

A Planet or the Moon Is Not Available on the Hand Control Although It Is Visible in the Sky

(NexStar 8/11 GPS, NexStar 5i/8i, NexStar 60/80/114/4)

Either the date/time/time zone is set incorrectly or the object is outside of the current Filter Limits. Refer to Chapter 5 for details on adjusting the date/time/time zone and Filter Limits.

Hand Control Compatibility Between NexStar models

(All models)

Although all NexStar hand controls (HC) are identical in appearance, they are not universally compatible. Following are the three families of NexStar telescopes. Models in the same family use the same hand control. Models in one family cannot use the HC from another family. Families are listed in the order they were introduced on the market. It should also be noted that some families have had upgraded HCs to fix problems and add features – this is discussed in Chapter 3:

- NexStar 5 and 8;
- NexStar 60, 80, 114 and 4 – note that the Tasco StarGuide series is compatible with this family;
- NexStar 8/11 GPS and NexStar 5i/8i – the NexStar 5i/8i require hand control version 2.2 or higher; versions 1.2 and 1.6 only work with the NexStar 8/11 GPS.

Fixing Incorrect Local Time on the Hand Control

(NexStar 8/11 GPS, NexStar 5i/8i with GPS Module)

If you check the local time on the NexStar hand control and find that it is wrong (in hours), there is a simple fix. The scope gets its time from the GPS satellites, an extremely accurate source, but the satellites report Universal Time. It is up to you to tell the scope what time zone it is located in. To do this, press the Menu button, scroll down to Setup Time–Site and select the correct time zone.

Potential Fix for Hand Control Misbehavior

(NexStar 8/11 GPS, NexStar 5i/8i, NexStar 60/80/114/4)

If your hand control starts to act strangely or won't run through the alignment procedure correctly, a potential solution is to reset the hand control to the factory settings. For the NexStar 8/11 GPS and 5i/8i, use the Factory Setting option on the Utilities menu to reset the hand control to its default settings. This has even been reported to work on new scopes and replacement hand controls received directly from Celestron. Note that this will erase any stored settings such as PEC and backlash compensation. Further information on the Factory Setting function can be found in Chapter 5.

For the NexStar 60/80/114/4, reset the Model Select on the Menu button. This has the effect of resetting all options in the hand control. For further information on the Model Select setting, refer to Chapter 5, **especially** if you have the original GT hand control.

When NexStar 8/11 GPS and NexStar 5i/8i Don't Seem to Come Close to the Alignment Stars During GPS Alignment

(NexStar 8/11 GPS, NexStar 5i/8i with GPS Module)

The GPS models use an electronic compass to determine magnetic north. In most areas, magnetic north is significantly different from true north – a phenomenon called magnetic declination or deviation. In some locations this deviation can be as much as 20 degrees! GPS hand control version 1.6 and higher includes a calibration function that automatically corrects for this deviation. Refer to Chapter 5 to learn about this calibration.

If you have a NexStar 8/11 GPS with hand control version 1.2, you can avoid this problem by forcing the GPS alignment to use true north instead. After you start the GPS alignment, the telescope will level the tube and begin to turn towards north. After the tube is level and while the scope is still locating north, press the Undo button. Then, release the azimuth clutch, point towards true north yourself (refer to Polaris when possible), set the clutch again, and then press Enter to resume the GPS alignment. The scope will use your north direction and move on accurately to the first alignment star.

Keep in mind that regardless where the telescope initially points for the alignment stars, the only important factor for alignment accuracy is the precision with which you center the alignment stars. The compass inaccuracies will only affect the telescope's ability to assist you in locating those stars.

Problems Obtaining GPS Link

(NexStar 8/11 GPS, NexStar 5i/8i with GPS Module)

Generally the NexStar 8/11 GPS and NexStar 5i/8i with GPS module will successfully link with the GPS satellites in less than one minute. Occasionally the link may take a little longer, particularly if you have moved your telescope a great distance since the last use – hundreds of miles for example – or haven't used your scope in a long time. If it seems you have waited long enough and the scope still has not achieved a link, the most common problem is an obstruction between the telescope and the GPS satellites. For example, it is generally not possible to get a link inside a building. When outside, a building or a large tree can obstruct the signals from the satellites.

One other potential problem is a slight incompatibility between hand control version 2.2 and some of the earliest NexStar 8/11 GPS telescopes. The hand control might query the GPS module before it is initialized. If this happens, you will not get a GPS link even if you wait for hours.

If you have a NexStar 8/11 GPS with hand control 2.2 and experience this problem, there is a workaround. After waiting a minute or two, if the scope doesn't link, press Undo until you are back at the initial startup prompt and start over again. If it doesn't link immediately, press Undo and restart the alignment again. Normally it will get the link the first time, unless you were just being impatient.

No Response 16/17 Error

(NexStar 8/11 GPS, NexStar 5i/8i)

This error indicates that the hand control is unable to communicate with the motor control processors located inside the mount. As discussed in the tip on power problems, low power sometimes causes this error. If the problem doesn't seem to be power-related, there may be a loose connector inside the telescope.

For the NexStar 8/11 GPS, start by disconnecting the power from the telescope. Using a $\frac{3}{16}$-inch hex wrench, remove the two bolts securing the handle to the left fork arm. Using a $\frac{9}{64}$-inch or 3.5mm hex wrench, remove the plastic covering on the left fork arm. Remove and reseat all connector cables attached to the small motor control circuit board. Be **very** careful not to misalign the pins and the connectors or damage may result. Replace the cover and the handle and test the telescope.

If you still experience problems, there may be a connector loose under the panel on the base of the scope. Using a $\frac{9}{64}$-inch or 3.5mm hex wrench, remove the five bolts on the top of the base panel. In a well-lit space, slowly raise the panel, being careful not to pull any of the cables loose. With the panel raised about 2 inches, reseat all connector cables you can easily reach. Reseat the panel and secure the five bolts.

For the NexStar 5i/8i, open the battery cover and remove the four screws securing the base cover. Carefully lift the base cover, noting there is a cable running from the battery holder to the base. Remove and reseat all connector cables attached to the circuit boards. Be **very** careful not to misalign the pins and the connectors or damage may result. Replace the cover and test the telescope.

If this does not resolve your problem, it is time to call Celestron.

Errors in Messier Objects in Hand Control Database

(NexStar 5/8, Original NexStar 60/80/114/4 GT, Version 1.2 of NexStar 8/11 GPS)

The hand controls for the NexStar 5/8 and original 60/80/114/4 contain three Messier objects with mistaken coordinates – M2, M10, and M110. The correct coordinates I give below are Epoch 2000 – referenced to the year 2000. As a workaround, use the Goto RA–Dec command on the menu to GoTo these objects. Note that the new GT hand control has corrected these errors for the 60/80/114/4 models. The NexStar 5i/8i has no coordinate errors in Messier objects. Other than the error for M2 noted below for the NexStar 8/11 GPS (hand control version 1.2 only), the NexStar 8/11 GPS has no coordinate errors in Messier objects.

NexStar 5/8 and Original GT

M2

NexStar: 21^h 33.5^m $+0^d$ 48.8^m
Epoch 2000: 21^h 33.5^m -0^d 49^m

M10

NexStar: 16^h 59.1^m -52^d 43.1^m
Epoch 2000: 16^h 57.1^m -4^d 6^m

M110

NexStar: 11^h 57.6^m $+53^d$ 23.0^m (this is a duplicate coordinate set of M109)
Epoch 2000: 0^h 40.4^m $+41^d$ 41^m

Hand Control Version 1.2 for the NexStar 8/11 GPS

M2

NexStar: 21^h 33.5^m $+0^d$ 35^m
Epoch2000: 21^h 33.5^m -0^d 49^m

Unable to Read Scrolling Text on the LCD Display in Cold Weather

(All Models)

LCD panels do not operate well in cold weather. Refer to the section "Cold Weather Gear for Your Scope" of Chapter 8 for solutions.

Tripod Feet Fall Out on the Heavy Duty Tripod

(Tripod 93499 and 93501)

Several owners have reported that one or more of the metal feet has come loose on their Heavy Duty Tripod. Losing one of these in the field would be a shame, but fortunately there is an easy fix. Several companies make a metal-based epoxy; JB Weld and Poxy Weld are two fine examples. Look for such products in an automotive parts store in your area.

NexStar 8/11 GPS

Correct Installation of the 9×50 Finderscope

You wouldn't expect the installation of a finderscope to be much trouble, but the 9×50 finderscope supplied with the NexStar 8/11 GPS is a little tricky. I had real problems with my 9×50 the first few nights; it just would not stay aligned. I then realized that the large groove or channel towards the back is to be ignored; don't put the rubber O-ring in it as intuition would suggest. Rather, push the O-ring further to the back (eyepiece end) so that it seats firmly in the back ring of the mounting bracket when the screws in the front ring are set in the small groove towards the front of the finderscope.

Mounting the NexStar 8/11 GPS on the Tripod

Due to the weight, mounting the NexStar 8/11 GPS on the tripod seems daunting the first few tries. Practice will make it easier. One potential problem is if the center pin on top of the tripod is set too high or low. The top of the pin should be located about $\frac{3}{8}$ inch (9 mm) above the top of the tripod. Loosen the nut securing the threaded rod to the bottom of the tripod head to make adjustments.

After you have the scope seated on the tripod, you must align the boltholes. Engage the azimuth clutch to allow use of a fork arm to rotate the base. Kneel down and look for the metal rib between the power connector and the power switch. When this is centered above any of the tripod legs the boltholes are aligned. Insert the first bolt, but do not tighten it. It is likely that you will need to rotate the base slightly to start the other two bolts. After all three bolts are started, tighten them all by hand.

If you still have difficultly, Starizona sells a product they call the Landing Pad. The Landing Pad guides the scope base to the center of the tripod head, taking the guesswork out of finding the center pin. It also has three recesses that the base locks into when rotated – these align the boltholes.

Rumbling Sound When Rotating Left/Right with Clutch Released

This is generally caused by dirty roller bearings. Refer to Chapter 12 for instructions on cleaning them.

Clicking Sound When Slewing Left/Right

A few owners have found that their telescope clicks slightly when slewing left and right. Misrouted cables in the base of the scope can cause this disturbing sound. To fix the problem, use a $\frac{9}{64}$-inch or 3.5mm hex wrench to remove the five bolts on the top of the base panel. In a well-lit space, slowly raise the panel, being careful not to pull any of the cables loose. All cables should be routed so that they do not make contact with the slipring circuit boards. These circuit boards are horizontally mounted in the center of the base; one is stationary, the other rotates with the scope. Reroute the cables so that they are lower than the bottom circuit board. Test your results by rotating the telescope 360° in azimuth. Reseat the panel and secure the five bolts.

Loose Secondary Mirror on the NexStar 8/11 GPS

The Fastar secondary assembly holds the secondary mirror in place on the corrector plate. Insure the large lockring at the front of the scope is tightened by hand.

Using Hand Control Version 2.2 or Higher with an Older NexStar 8/11 GPS

The newest NexStar 8/11 GPS telescopes come with hand control version 2.2 or higher. Some elect to purchase the separately available version 2.2 hand control – sold as the computerized hand control upgrade for the NexStar 5i/8i – as a spare or replacement for the NexStar 8/11 GPS. This version requires an index mark on the fork arm indicating the position where the optical tube is perpendicular to the fork arm. If you purchase or receive a version 2.2 hand control for an older scope without the index mark, you will need to make one.

Affix tape to both sides of the joint between the fork arm and the rail supports as shown in figure 4-2 in the hand control version 2.2 addendum supplied by Celestron (the addendum is also available from Celestron's download site). Carefully level the tripod and mount the scope. Then perform a GPS alignment. Insure the alignment is accurate (good GoTo performance), and then perform the Compass calibration and the Alt Sensor (the new name for the level switch) calibration. Next, **turn tracking off** via the menu. Then on the Alt Sensor menu, use the option to move to the alt sensor. This actually moves to the Alt = 0 location. Use a ruler and a pen to draw a line across the two pieces of tape to indicate this Alt = 0 location.

NexStar 8/11 GPS Vibrates or Hops While Tracking Objects

Some of the first NexStar 11 GPS scopes to reach owners presented a problem with vibration (which could actually be felt at the diagonal) and a slight "bump" in tracking (the object in the eyepiece would drift from the center and then return). Certain versions of the motor control firmware resurfaced the "bump" problem. If your NexStar 8/11 GPS has such a problem you should contact Celestron Tech Support (support@celestron.com) to learn how to update your motor control firmware.

NexStar 8/11 GPS Loops Over the Top When Locating Level During GPS Alignment

Almost anyone who owns an early version of the NexStar GPS has watched their scope loop all the way over while finding level and wondered just how silly a computer can be! While generally amusing, it can be horrifying if you happen to have any accessories that are damaged by the unusual positions of the optical tube assembly (OTA) during these maneuvers. This problem was addressed starting in hand control version 1.6, but if you have version 1.2 it is easy to prevent. Before starting the alignment, insure the OTA is pointed with the front about 10° below level. The level position is located only as the front of the OTA is moving up.

Version 1.6 tries to prevent this by dipping slightly when it first starts the alignment. Providing you start with the OTA about level, it will not loop over when seeking the level position.

Left and Right Buttons Reversed or Unable to Find North after Replacement of Motor Control Board

If you have a very early NexStar 8/11 GPS, you may be required to send your motor control (MC) board to Celestron for replacement. After contacting Celestron Tech Support, they will send you the required instructions.

After you receive the replacement and install it, you may find that you have problems with the left/right arrow buttons or the scope may not be able to find north. The scope may point south instead of north during the GPS Alignment routine or continue around in circles. Most likely the MC board you replaced was the oldest version and one of the cables you reconnected has a short extension cable that reverses the azimuth motor leads. Simply remove the extension cable and connect the main cable directly to the MC board.

Some of the older boards that require the extension cable have been updated as well. So if you have the problem after removing the extension cable, reinstall it and try again. If you have the problem and your scope had no jumper to begin with, call Celestron.

It is also possible to experience problems ranging from no movement at all, to inoperative compass or GPS functions. Check to be sure all of the cables to the MC board are correctly oriented, all the silver pins on the board are in a corresponding hole in a cable connector, and all the connectors are fully seated.

NexStar 5/8

Locations (Longitude And Latitude) Not Stored Correctly on NexStar 5/8

The original NexStar 5/8 offers the ability to store up to 10 observing locations. Unfortunately there is a problem that causes new entries to corrupt the adjacent three entries. Thus, you can only use entries 0, 4, 8 or 1, 5, 9.

Runaway Slews and Hand Control Lockup on Original NexStar 5

(NexStar 5)

Some early models of the NexStar 5 had problems with runaway slewing (the scope takes off during a GoTo and never stops) and hand control lockup, especially when using a computer to control the scope. Some have claimed that low power (weak batteries) or a loose power connector causes most runaways, but others feel it is more likely a problem with the firmware in the hand control. It is known that newer versions of the firmware resolved the problem with computer-control lockups as well as most runaway problems.

If you believe you are experiencing these problems, contact Celestron for assistance.

NexStar 60/80/114/4

Attaching Accessories to the Rear Port of the NexStar 4

Any accessory designed to fit the rear port of the Celestron C90 or the Meade ETX will fit the rear port of the NexStar 4. Some of the accessories, discussed in Chapters 8 and 11, include:

- focal reducers;
- T-Adapters;
- 1.25 visual backs (not discussed elsewhere in the book); Meade part number 07812;
- thread-on diagonals – but note that not all eyepieces can reach focus with a diagonal attached to the rear port;
- Celestron's Large Accessory Ring (LAR) – provides a standard SCT thread allowing SCT accessories to be attached as well.

Misplaced Secondary Baffle on NexStar 4

Some of the earliest NexStar 4s to reach owners suffered from a misplaced secondary baffle. The secondary baffle is a tube that is mounted on the inside of the corrector lens at the front of the telescope. It should be centered over the secondary mirror spot on the inside of the corrector. On some NexStar 4s, the glue that holds the baffle in place seems to have melted and the baffle shifted from the correct position. While Celestron fixed the problem soon after the initial release, one of these units will still occasionally surface, especially on the used telescope market. If you are the unhappy owner of a NexStar 4 with a slipped baffle, contact Celestron for repair.

FOV Moves Up/Down When Slewing Left/Right

If you find that the objects in the field of view on your NexStar move up (the view seems to hop) when you start slewing left or right, the most likely cause is a loose azimuth axis. To correctly adjust the azimuth axis, refer to Chapter 12.

NexStar 80 (Short-Tube 80mm F/5) Tune-Up

(Also applies to many other Chinese-made refractors such as the short-tube 90 and 120mm models)

By Bob Berta

Disclaimer: The following are all easy user fixes but of course they may cancel your warranty and the author cannot be responsible for any problems that arise. But the small risk is worth it; the improvement in performance can be dramatic!

First, some scopes exhibit a bit of pinched optics defect. Test your scope by pointing at a bright star and slightly defocusing the image. You should see a round image that resembles a bull's-eye. If the image is not round but shows three semi-corners you have pinched optics. The objective lens is actually two air-spaced lenses (good) and the spacing material is three thin shims between the two elements. A retaining ring that is a little too tight causes pinched optics. To fix this, slide off the dew cap/lens hood – it pulls straight out. Look for a threaded plastic retaining ring just in front of the lens. Simply unscrew this ring a very small amount and see if the problem goes away. Note that this can be temperature-sensitive – you may not have pinched optics at warmer temperatures, but in very cold temperatures it may appear – so it might be necessary to adjust the ring based on temperature.

The next fix requires you to actually remove the lenses from the holding cell. First, remove the scope from the tube ring on the fork arm. Then, unscrew and remove the retaining ring described above. Be sure the scope is aiming upwards – you don't want the lenses to fall to the floor! Take a clean, lint-free cloth (a handkerchief is good) and place it over your hand. Place your cloth-covered hand over the end of the scope and point it front-down. If the lenses don't slide out into your hand, keep your hand in place and gently tap the scope, hand and all, on a hard surface to make them slip loose. Remember, the lens assembly has two elements. You could separate those two elements from each other, but it is best to keep them together. **Warning:** Make sure you note the orientation of the lens pair; you don't want to put the lens in backwards! While you have the lenses in hand, use a wide-tip, black, permanent marker and blacken the edges of the optics. Don't use paint, as this will often make the lenses too tight for the cell. Set the lenses aside on the cloth and go to the next step.

Look inside the scope from the now open front of the tube. You will see several shiny screws, the end of the focusing tube, and other bits of bare metal. Using flat black model paint, paint anything shiny with a long-handled brush. You might consider removing the focusing assembly by removing the three large chrome screws.

NexStar 80 (Short-Tube 80mm F/5) Tune-Up *(continued)*

since they thread into thick aluminum but don't remove the three screws that hold the plastic lens cell assembly to the front of the tube. Those three screws thread into soft, thin aluminum and strip easily. However, if you do damage the threading, drill out the holes to $\frac{1}{8}$" and replace the screws with pop rivets (don't forget to paint them flat black).

After the paint dries you may reassemble the scope. When reinstalling the doublet lens, don't just "drop" it into the lens cell. It may not seat properly and it is possible the thin metal shims will shift. Rather, place the lens setup on top of your cloth-covered hand and invert the scope tube. Use your hand to push the lens up into the scope keeping it square and making sure it seats fully. Replace the retaining ring ensuring you don't over-tighten to avoid pinched optics. It is a good idea to press down around the edges of the lens with the cloth to insure it is well seated.

The last fix is to remove any excess play from the focuser assembly. Look closely at the very top of the focuser near the focus-lock thumbscrew – you will see two tiny holes. These contain hex screws. The focusing assembly rides on plastic rails. The upper one is adjustable via these two hex screws. Loosen the focus-lock thumbscrew and try to wiggle the rear of the focusing tube. If it moves up and down relative to the scope, you can improve the situation by tightening the two hex screws. Adjust both of the screws to get the upper rail to ride smoothly at both ends of the focuser travel.

Once completed your scope will have no pinched optics, reduced ghost images around very bright objects like the Moon, reduced flare, increased contrast, increased sharpness, and smoother focuser action. Quite a payback for very little work.

NexStar 60/80/114/4 with Original GT Hand Control

These tips also apply to the Tasco StarGuide telescopes.

Original GT Model Hand Control: Known Bugs

The original GT hand control – the GoTo hand control for the NexStar 60/80/114/4 and the only hand control for the Tasco StarGuide models – has several known flaws, chief among them being:

- No official RS-232 support – however, several programs are available for PC and palmtop computer control of the original GT hand control.
- Won't save observing locations (longitude/latitude) in the Eastern Hemisphere – it is necessary to re-enter the longitude and latitude during every alignment for locations in the Eastern Hemisphere.

- Won't save backlash settings.
- Cordwrap feature not functional.
- Occasional lockups and runaway slews.
- Very touchy keypad that often leads to double key entries.
- Light Control feature not functional.
- Tracking Rate defaults to Solar.

Celestron released an updated hand control with all telescopes manufactured after December 2001. All of the above issues have been corrected. Contact Celestron Tech Support via their web site (http://www.celestron.com) if you have the original GT hand control. You can determine if you have the original model by any of the bugs above or by the lack of the Two Star Alignment option that was added to the new GT hand control. If you cannot get an updated hand control (Tasco did not release an updated version), the following tips will help you enjoy your telescope.

Avoiding Hand Control Lockup

The original GT hand control on NexStar telescopes can be a little temperamental at times. Occasionally it will not respond to the buttons, particularly right after finishing a slew. A little patience goes a long way in this case. After a slew, it sometimes takes as long as 20–30 seconds before the hand control is ready to respond again. Try not to get in a hurry and you will find things go smoothly. When the hand control is still unresponsive after a short wait, try pressing the direction buttons (the arrows) or the Undo button – this usually wakes it up.

Third Star Align Feature

The Third Star Align feature (now called Re-Align) on the original GT NexStars is reported to be nonfunctional, but actually it does work. The problem is that the instructions in the manual are incorrect or at least difficult to follow. Like the description in Chapter 4, this feature allows you to replace either of the current alignment stars with any of the 40 alignment stars in the Named Stars list. This can help to improve accuracy in the area of the new alignment star.

Here is the correct procedure:

- Press the LIST key, select Named Stars, scroll to your desired alignment star, and then press the ENTER key to slew to that star.
- Center the star in the eyepiece using the direction keys.
- Press the UP button (the 6) then the DOWN button (the 9) to reselect the star in the Named Stars list.
- Press the ALIGN key.
- You will be prompted to choose the original alignment star to be replaced, use the UP or DOWN key (6 and 9) to select the star you wish to replace.
- Press the ENTER key.

Sounds tough, but after you do it a few times it will be second nature.

Simulating a Two-Star Alignment

The NexStar 60/80/114/4 with the original GT hand control does not offer the option of a two-star alignment procedure. The two-star alignment feature on the bigger NexStars and the new GT control allows you to choose the alignment stars yourself rather than having the auto-align do it for you. As discussed in Chapter 4, selection of alignment stars is important for improved GoTo performance.

Hank Williams offers the following suggestion. Perform the Auto Align, simply accepting the location of the two stars without bothering to center. Then, after the alignment is complete, use the Third Star Align feature (above) to replace the two Auto Align stars with two of your choosing. Naturally you would be careful to center the stars during the Third Star Align procedure and be certain to replace **both** of the original Auto Align stars.

Preventing Lockup When Using Computer Control

If you connect a computer to the original GT hand control, you will soon find that it is very easy to lock up the hand control when sending GoTo commands from the computer. The solution is simple; remembering to use the solution might be more difficult.

After each GoTo, use the arrow buttons on the hand control to move the scope manually. Generally you will do this anyway to center the object in the eyepiece, but should you change your mind and decide to immediately GoTo another object, be sure to press an arrow button anyway to prevent lockup.

Conclusion

I hope these tips will help you to get the most from your NexStar. In the Odds and Ends section of the NexStar Resource Site – http://www.NexStarSite.com – you will find many of these and newer tips introduced after this book was published. Suggest your favorite NexStar workaround or improvement by contacting me at swanson.michael@usa.net.

Appendix A

Internet Resources

NexStar-Specific Web Sites

Site name	Description	Web site address
NexStar Resource Site	The author's web site	www.NexStarSite.com
NexStar FAQ Site	Articles and tips for NexStar telescopes	www.grcooperjr.com/faq.htm
Yahoo NexStar Group	Discussion group for all models of NexStar telescopes	groups.yahoo.com/group/NexStar
Yahoo NexStarGPS Group	Discussion group for the GPS models of NexStar telescopes	groups.yahoo.com/group/NexStarGPS
NexStar 50 Club	Web site for the NexStar 50 observer clubs – awards are given for a general list of objects and observation of lunar features	www.NexStarSite.com/nexstar50club.htm
NexStar Image Gallery	A collection of hundreds of images taken with all models of NexStar telescopes	www.buyastrostuff.com/dons
Matthias Bopp's Site	A unique collection of technical information and tips for NexStar telescopes	www.dd1us.de
The Unofficial NexStar 11 Support Site	Dave Cole's web site featuring tips and resources for the NexStar 8/11 GPS telescopes	www.nexstar11.com
Celestron's Web Site	Official web site of Celestron, manufacturer of NexStar telescopes	www.celestron.com
Celestron's Download Site	Product manuals and other downloads	www.celestron.com/downloads/index.htm

General-Astronomy Web Sites

Site name	Description	Web site address
American Association of Amateur Astronomers	Web site of the Association – general astronomy information	www.corvus.com
Astronomical League	Web site of the League – an organization for amateur astronomers, general astronomy information, observation programs and awards	www.astroleague.org
Astronomy Magazine	Web site of *Astronomy* magazine – news and general information for amateur astronomers	www.astronomy.com
Cloudy Nights Telescope Reviews	Equipment reviews submitted by amateur astronomers	www.cloudynights.com
NASA Web Site	Official web site of the United States' National Aeronautics and Space Administration	www.nasa.gov
ScopeReviews.com	Equipment reviews by Ed Ting	www.scopereviews.com
Sky and Telescope	Web site of *Sky and Telescope* magazine – news and general information for amateur astronomers	skyandtelescope.com
SkyMaps	Free monthly sky charts	www.skymaps.com
Students for the Exploration and Development of Space	Great source of information on solar system, deep sky objects, and general astronomy	seds.lpl.arizona.edu

Vendor and Manufacturer Web Sites

Company	Products – notes	Web site address
Adirondack Video Astronomy	CCD and video cameras, automated focusing software, imaging accessories	www.astrovid.com
Adorama	Telescopes, accessories	www.adoramaphoto.com
Alpine Astronomical	The Baader Bracket, Baader astrophotography adapters, other accessories	www.alpineastro.com
Anacortes Telescopes	Telescopes, accessories	www.buytelescopes.com
Apogee Instruments	CCD astronomy cameras	www.ccd.com
Apogee, Incorporated	Telescopes, accessories	www.apogeeinc.com
AstroMart	Used equipment; you can both buy and sell	www.astromart.com

Company	Products – notes	Web site address
Astronomics	Telescopes, accessories	www.astronomics.com
Astronomy-Mall	A short listing of astronomy vendors and free classified ads for astronomy equipment	astronomy-mall.com
Astro-Physics	Diagonals, Baader Solar Film, Baader Color Filters, Baader Skyglow Filter, portable piers	www.astro-physics.com
Baader Planetarium	Telescopes, the Baader Bracket, filters, astrophotography accessories, other accessories	www.baader-planetarium.de
BuyAstroStuff.com	Ray's Bracket, observing chairs	www.buyastrostuff.com
Campmor	Outdoor equipment supplier	www.campmor.com
Cases and Covers	You guessed it – telescope cases and covers	casesandcovers.com
Coleman	Outdoor equipment supplier	www.coleman.com
Coronado	H-Alpha solar filters	www.coronadofilters.com
GRABBER Warmers	Chemical hand warmers	www.grabberwarmers.com
Hands On Optics	Telescopes, accessories	www.handsonoptics.com
ITE Telescopes	CCD and video cameras, other accessories	www.itetelescopes.com
Jim's Mobile, Inc. (JMI)	Telescope accessories	www.jimsmobile.com
Kendrick Astro Instruments	Dew prevention systems, other accessories	www.kendrick-ai.com
Losmandy	Rails, counterweights, mounting accessories	www.losmandy.com
Morrow Technical Services	Bob's Knobs (no-tools collimation screws for SCTs), no-tools tripod bolts for the N8/11GPS	www.bobsknobs.com
Optec	Focal reducers (telecompressors)	www.optecinc.com/astronomy
Orion Telescopes and Binoculars	Telescopes, accessories	www.telescope.com
Pelican Cases	Telescope cases	www.pelican.com
Peterson Engineering	EyeOpener 2" adapter for SCTs	www.peterson-web.com
REI	Outdoor equipment supplier	www.rei.com
S. King Company	A collection of 12V d.c. convenience equipment	www.skingco.com
SAC Imaging	Video astronomy cameras	www.sac-imaging.com
Santa Barbara Instrument Group (SBIG)	CCD astronomy cameras	www.sbig.com
ScopeGuard Cases	Telescope cases	www.scopeguard.com
ScopeStuff	Accessories	www.scopestuff.com

Company	Products – notes	Web site address
ScopeTronix	Digital camera adapters, other accessories	www.scopetronix.com
SightAndSoundShop	Telescopes, accessories	www.sightandsound shop.com
SKB	Golf cases suitable for some telescopes and tripods	www.skbcases.com
Starizona	Telescopes, accessories	www.starizona.com
Starlight Xpress	CCD astronomy cameras	www.starlight-xpress.co.uk
Tele Vue	Telescope, eyepiece, and accessory manufacturer	www.televue.com
The Telescope Warehouse	Unique astronomy accessories	www.astronomy-mall.com/telescope-ware house
Thousand Oaks	Filters	www.thousandoaks optical.com
University Optics	Eyepieces, Barlow lenses	www.universityoptics.com
Van Slyke Engineering	Zero-image-shift focusers, off-axis guiders	www.observatory.org
Warehouse Photo	T-mounts and various camera adapters	www.warehousephoto.com/lenses&.htm
William Optics	Telescopes, accessories	www.william-optics.com

Appendix B
Objects in the NexStar Hand Control

The objects in the NexStar database are organized by several categories. Several of the categories are the well-known catalogs discussed in Chapters 2 and 6.

Named Stars (Alignment Stars)

The brightest stars in the sky are listed, primarily for use as alignment stars.

Name	RA	Dec	Mag	Constellation
Achernar	01^h 37^m 43^s	–57°14'	0.6	Eri
Acrux	12^h 26^m 36^s	–63°05'	1.6	Cru
Albireo	19^h 30^m 43^s	+27°58'	3.2	Cyg
Aldebaran	04^h 35^m 55^s	+16°31'	1.1	Tau
Algenib	00^h 13^m 14^s	+15°11'	2.9	Peg
Alpha Centauri	14^h 39^m 36^s	–60°50'	0.0	Cen
Alphard	09^h 27^m 35^s	–08°39'	2.2	Hya
Alphecca	15^h 34^m 41^s	+26°43'	2.3	CrB
Alpheratz	00^h 08^m 23^s	+29°06'	2.1	And
Altair	19^h 50^m 47^s	+08°52'	0.9	Aql
Antares	16^h 29^m 24^s	–26°26'	1.1	Sco
Arcturus	14^h 15^m 40^s	+19°13'	0.2	Boo
Betelgeuse	05^h 55^m 10^s	+07°24'	0.6	Ori
Canopus	06^h 23^m 57^s	–52°42'	–0.9	Car
Capella	05^h 16^m 41^s	+46°00'	0.2	Aur
Caph	00^h 09^m 11^s	+59°09'	2.4	Cas
Castor	07^h 34^m 36^s	+31°53'	1.6	Gem
Deneb	20^h 41^m 26^s	+45°17'	1.3	Cyg
Denebola	11^h 49^m 04^s	+14°34'	2.2	Leo
Dubhe	11^h 03^m 44^s	+61°45'	2	Uma
Fomalhaut	22^h 57^m 39^s	–29°37'	1.3	PsA
Hadar	14^h 03^m 50^s	–60°23'	0.9	Cen
Hamal	02^h 07^m 10^s	+23°28'	2.2	Ari
Mimosa	12^h 47^m 43^s	–59°41'	1.5	Cru
Mirach	01^h 09^m 44^s	+35°37'	2.4	And

Mirfak	03^h 24^m 19^s	+49°52'	1.9	Per
Mizar	13^h 23^m 56^s	+54°55'	2.4	Uma
Navi	00^h 56^m 43^s	+60°43'	2.8	Cas
Peacock	20^h 25^m 39^s	−56°44'	2.1	Pav
Polaris	02^h 31^m 50^s	+89°16'	2.1	Umi
Pollux	07^h 45^m 19^s	+28°02'	1.2	Gem
Procyon	07^h 39^m 18^s	+05°14'	0.5	Cmi
Rasalhague	17^h 34^m 56^s	+12°34'	2.1	Oph
Regulus	10^h 08^m 22^s	+11°58'	1.3	Leo
Rigel	05^h 14^m 32^s	−08°12'	0.3	Ori
Scheat	23^h 3^m 46^s	+28°05'	2.6	Peg
Sirius	06^h 45^m 9^s	−16°42'	−1.6	Cma
Spica	13^h 25^m 12^s	−11°10'	1.2	Vir
Suhail	09^h 08^m 00^s	−43°26'	2.2	Vel
Vega	18^h 36^m 56^s	+38°47'	0.1	Lyr

Double Stars

There are 55 double stars in the NexStar database. Additional double stars can be accessed with the **Star** button on the hand control. Lists of those stars (hundreds for the GT, thousands for the other NexStar models) are available in the Downloads section of my NexStar Resource Site – http://www.NexStarSite.com."

Name	Ra	Dec	Mag	Sep/Pos angle	Constellation
107 Aqr	23^h 46^m 01^s	−18°41'	5.8	7" / 143°	Aqr
145 Cma	07^h 16^m 36^s	−23°19'	4.8	27" / 65°	Cma
17 Cyg	19^h 46^m 26^s	+33°44'	5.0	26" / 73°	Cyg
19 Lyn	07^h 22^m 54^s	+55°17'	5.6	14.8" / 315°	Lyn
24 Com	12^h 35^m 08^s	+18°22'	5.2	20" / 271°	Com
30 Ari	02^h 37^m 00^s	+24°39'	6.6	39" / 274°	Ari
32 Eri	03^h 54^m 18^s	−02°57'	5.0	7" / 347°	Eri
35 Com	12^h 53^m 18^s	+21°15'	5.1	29" / Triple	Com
38 Gem	06^h 54^m 39^s	+13°11'	4.7	7"	Gem
54 Leo	10^h 55^m 37^s	+24°45'	4.5	6.8" / 110°	Leo
61 Cyg	21^h 06^m 36^s	+38°42'	5.4	30" / 148°	Cyg
94 Aqr	23^h 19^m 07^s	−13°27'	5.3	13" / 350°	Aqr
95 Her	18^h 01^m 30^s	+21°36'	5.1	6.5" / 258°	Her
Acamar	02^h 58^m 16^s	−40°18'	4.4	8" / 82°	Eri
Adhara	06^h 58^m 38^s	−28°58'	1.6	7.5" / 161°	Cma
Albireo	19^h 30^m 43^s	+27°58'	3.2	35" / 54°	Cyg
Algieba	10^h 19^m 58^s	+19°50'	2.6	4.4" / 124°	Leo
Algorab	12^h 29^m 52^s	−16°31'	3.1	24" / 214°	Crv
Almach	02^h 03^m 55^s	+42°20'	5.1	10" / 63°	And
Beta Mon	06^h 28^m 49^s	−07°02'	4.7	3" / 132°	Mon
Castor	07^h 34^m 36^s	+31°53'	1.6	1.8" / 171°	Gem
Cor Caroli	12^h 56^m 01^s	+38°19'	2.9	19" / 229°	CVn
Dabih	20^h 21^m 01^s	−14°47'	3.3	205" / 267°	Cap
Delta Cep	22^h 29^m 10^s	+58°25'	4.0	20" / 192°	Cep
Delta Ser	15^h 34^m 48^s	+10°32'	5.2	3.9" / 179°	Ser
Epsilon Boo	14^h 44^m 59^s	+27°05'	2.7	3" / 303°	Boo
Epsilon Lyr 1	18^h 44^m 20^s	+39°40'	6.0	2" / 353°	Lyr
Epsilon Lyr 2	18^h 44^m 23^s	+39°37'	4.5	2.2" / 80°	Lyr
Epsilon Peg	21^h 44^m 11^s	+09°53'	2.5	83" / 320°	Peg

Eta Cas	00^h 49^m 03^s	+57°49'	3.6	12" / 312°	Cas
Eta Puppis	07^h 34^m 20^s	−23°28'	5.9	10" / 117°	Pup
Gamma Aries	01^h 53^m 30^s	+19°19'	4.8	8" / 360°	Ari
Gamma Cet	02^h 43^m 18^s	+03°14'	3.6	2.7" / 297°	Cet
Graffias	16^h 05^m 26^s	−19°48'	2.9	14" / 21°	Sco
Iota Cancer	08^h 46^m 42^s	+28°46'	4.2	31" / 307°	Cnc
Kappa Boo	14^h 13^m 29^s	+51°47'	4.6	13" / 236°	Boo
Kappa Puppis	07^h 38^m 48^s	−26°48'	4.6	10" / 318°	Pup
Lamda Aries	01^h 57^m 54^s	+23°36'	4.8	38" / 47°	Ari
Mintaka	05^h 32^m 00^s	−00°18'	2.5	53"	Ori
Mizar	13^h 23^m 56^s	+54°55'	2.4	14" / 151°	Uma
Nu Dra	17^h 32^m 11^s	+55°11'	5.0	62" / 312°	Dra
Omicron Cap	20^h 29^m 54^s	−18°35'	6.1	19" / 239°	Cap
Polaris	02^h 31^m 50^s	+89°16'	2.1	18" / 218°	Umi
Porrima	12^h 41^m 40^s	−01°27'	2.9	3" / 287°	Vir
Psi Piscium	01^h 05^m 42^m	+21°28'	5.5	30" / 159°	Psc
Rasalgethi	17^h 14^m 39^s	+14°24'	3.5	4.6" / 107°	Her
Rigel	05^h 14^m 30^s	−08°12'	0.3	9" / 203°	Ori
Sigma Cas	23^h 59^m 01^s	+55°45'	4.9	3" / 326°	Cas
Tau 1 Aqr	22^h 47^m 43^s	−14°03'	5.7	23" / 121°	Aqr
Tegman	08^h 12^m 13^s	+17°39'	6.0	9" / 89°	Cnc
Theta 2 Cnc	08^h 26^m 48^s	+26°56'	6.3	5" / 217°	Cnc
Xi Boo	14^h 51^m 23^s	+19°06'	4.6	7" / 326°	Boo
Zeta Aqr	22^h 28^m 50^s	+00°01'	4.6	2" / 207°	Aqr
Zeta Lyr	18^h 44^m 46^s	+37°36'	4.3	44" / 150°	Lyr
Zeta Piscium	01^h 13^m 42^s	+07°35'	5.6	23" / 63°	Psc

Variable Stars

The hand control database includes a small selection of the most interesting variable stars.

Name	RA	Dec	Min mag	Max mag	Period in days	Constellation
Algol	03^h 08^m 10^s	+40°57'	2.2	3.5	2.9	Per
Beta Lyr	18^h 50^m 05^s	+33°22'	3.5	4.3	12.9	Lyr
Delta Cep	22^h 29^m 10^s	+58°25'	3.6	4.3	5.4	Cep
L Car	9^h 45^m 15^s	−62°30'	3.4	4.1	35	Car
Mira	02^h 19^m 21^s	−02°59'	3.4	9.2	332	Cet
R Car	09^h 32^m 15^s	−62°47'	4.6	9.6	309	Car
R Cen	14^h 16^m 34^s	−59°55'	5.9	10.7	546	Cen
R Cyg	19^h 36^m 26^s	+50°13'	7.5	13.9	426	Cyg
R Dra	16^h 32^m 39^s	+66°45'	7.6	12.4	246	Dra
R Hor	02^h 54^m 06^s	−50°52'	6.0	13.0	404	Hor
R Leo	09^h 47^m 34^s	+11°26'	5.8	10.0	312	Leo
R Tri	02^h 35^m 47^s	+34°41'	6.2	11.7	266	Tri
RR Lyr	19^h 25^m 27^s	+42°47'	7.3	8.1	0.6	Lyr
S Car	10^h 09^m 22^s	−61°33'	5.7	8.5	150	Car
SS Cyg	21^h 42^m 34^s	+42°26'	8.2	12.4	50	Cyg
T Cen	13^h 41^m 46^s	−33°36'	5.5	9.0	91	Cen
T Cep	21^h 09^m 32^s	+68°30'	6.0	10.3	388	Cep
U Gem	07^h 55^m 06^s	+22°40'	8.2	14.9	103	Gem
U Uma	10^h 44^m 38^s	+68°46'	7.5	13.0	302	Uma
Zeta Gem	07^h 04^m 07^s	+20°34'	3.7	4.3	10.2	Gem

Named Objects

A listing of some of the most spectacular deep sky objects.

Name	RA	Dec	Mag	Size	Type	Const	Comments
Andromeda Galaxy M31	00^h 42.7^m	+41°16'	3.5	178'	Gal	Andromeda	M31. The closest spiral galaxy to earth. Spanning 3° across, the Andromeda galaxy is the farthest object that can be seen with the naked eye. Distance = 2.8 million ly. Diameter = 180 000 ly.
Barnard's Galaxy NGC 6822	19^h 44.9^m	−14°48'	9.0	10.2'	Gal	Sagittarius	Irregular dwarf galaxy. Distance = 1.7 mil ly. Diameter = 10 000 ly.
Bear Paw Galaxy NGC 2537	08^h 13.2^m	+46°00'	11.7	1.7'	Gal	Lynx	Moderately bright, round, and well-resolved galaxy.
Beehive OC M44	08^h 40.1^m	+19°59'	3.1	95'	OC	Cancer	Open cluster containing over 200 stars, visible to the naked eye. Also called Praesepe and M44. Distance = 525 ly. Diameter = 40 ly.
Black-Eye Galaxy M64	12^h 56.7^m	+21°41'	8.5	9.3'	Gal	Coma Berenices	A spiral galaxy with a striking dark dust cloud near the central nucleus. Distance = 20 million ly.
Blinking Plan NGC 6826	19^h 44.8^m	+50°31'	10.0	2.3'	PN	Cygnus	Planetary nebula disc with 11th-magnitude central star.
Blue Planetary NGC 3918	11^h 50.3^m	−57°11'	8.0	0.2'	PN	Centaurus	Small, round planetary nebula. Blue in color with 7th-magnitude star in center.
Blue Snowball NGC 7662	23^h 25.9^m	+42°33'	9.0	2.2'	PN	Andromeda	A hazy blue planetary nebula estimated to be 1800 to 5600 ly away and 20 000 to 50 000 AU across.
Bode's Nebula M81	09^h 55.6^m	+69°04'	6.9	25.7'	Gal	Ursa Major	A massive galaxy with a luminosity equal to 20 billion suns and a total mass of approx 250 solar masses. Distance 7 million ly. Diameter = 36 000 ly.
Box Nebula NGC 6309	17^h 14.1^m	−12°55'	11.0	1.1'	PN	Ophiuchus	This planetary nebula has a greenish disk and appears rectangular in shape, hence its name. The Box Nebula is best seen using an OIII nebula filter.
Bubble Nebula NGC 7635	23^h 20.7^m	+61°12'		15'	SR	Cassiopeia	Remnant of a supernova, named for its gaseous arc

Name	RA	Dec	Mag	Size	Type	Const	Comments
							that resembles a large bubble. Distance = 11 000 ly.
Bug Nebula NGC 6302	17^h 13.7^m	–37°06'	13.0	0.8'	PN	Scorpius	Irregular-shaped planetary nebula.
Butterfly OC M6	17^h 40.1^m	–32°13'	4.2	15'	OC	Scorpius	Also called M6. Bright open cluster containing about 50 stars. Distance = 1300 ly. Diameter = 20 ly.
California Nebula NGC 1499	04^h 00.7^m	+36°37'		145'	Neb	Perseus	Very faint, large, and diffuse nebula resembling the shape of California. Best seen using a hydrogen-beta filter.
Cat's Eye Nebula NGC 6543	17^h 58.6^m	+66°38'	9.0	5.8'	PN	Draco	One of the brightest planetary nebulae with a central star nearly 100 times the luminosity of the Sun. Distance = 3200 ly. Diameter = 0.33 ly.
Christmas Tree NGC 2264	06^h 41.1^m	+09°53'	3.9	60'	OC	Monoceros	A very young open cluster, also called the Cone Nebula, with 150 members surrounded by dark nebulosity. Distance = 2600 ly. Diameter = 20 ly.
Cocoon Nebula IC 5146	21^h 53.4^m	+47°16'	7.2	12'	Neb	Cygnus	A faint open cluster surrounded by nebulosity that is located ESE of a large dark nebula – Barnard 168.
Crab Nebula M1	05^h 34.5^m	+22°01'	8.4	6'	SR	Taurus	Also called M1. Supernova remnant that was recorded by the Chinese in the year 1054. At its brightest, it was visible even in the daytime and cast a shadow in the nighttime. Distance = 6500 ly. Diameter = 10 ly.
Crescent Nebula NGC 6888	20^h 12^m	+38°21'		20'		Cygnus	A faint, very large nebular with a double star visible inside. Best seen with an OIII filter.
Deer Lick Galaxy NGC 7331	22^h 37.1^m	+34°25'	9.5	10.7'	Gal	Pegasus	A fairly bright elongated galaxy with four small companion galaxies surrounding it.
Double Cluster	2^h 19^m	+57°09'	4.0	30'	OC	Perseus	Two open clusters visible

Name	RA	Dec	Mag	Size	Type	Const	Comments
NGC 869 / 884							to the naked eye, each with a diameter of about 70 ly. Both clusters are very young. NGC 884 is 8500 ly in distance and 11.5 million years old. NGC 869 is 7000 ly away and 6.4 million years old.
Dumbbell Nebula M27	19^h 59.6^m	+22°43'	8.1	15.2	PN	Vulpecula	One of the closest planetary nebulae to earth. Expanding at a speed of 27 km/s, it can be estimated to be about 48 000 years old. Distance = ~750 ly. Diameter = 2 ly.
Eagle Nebula M16	18^h 18.8^m	−13°47'	6.0	35'	Neb	Serpens	A large open cluster surrounded by a diffuse nebula. A very active region of star formation. Distance = ~8000 ly. Diameter = 70 ly.
Eight Burst Nebula NGC 3132	10^h 07.0^m	−40°26'	8.0	0.8'	PN	Vela	Bright central star. Named after its multi-ring structure visible in photographs. Distance = 2500 ly. Diameter = 0.7 ly.
Eskimo Nebula NGC 2392	07^h 29.2^m	+20°55'	10.0	0.7'	PN	Gemini	Bright inner ring separated by a dark gap from another faint outer ring. Distance = approx. 2500 ly.
Ghost Of Jupiter NGC 3242	10^h 24.8^m	−18°38'	9.0	20.8'	PN	Hydra	Planetary nebula with an elongated inner ring. Distance = ~2600 ly. Diameter = 0.5 ly.
Helix Nebula NGC 7293	22^h 29.6^m	−20°48'		12.8'	PN	Aquarius	The largest and closest planetary nebula with an angular diameter of half the full Moon. Distance = ~300 ly. Diameter = 1.75 ly.
Hercules GC M13	16^h 41.7^m	+36°28'	5.9	16.6'	GC	Hercules	Globular cluster visible to the naked eye. The cluster contains about 1 million stars and is estimated to be 10 billion years old. Distance = 25 000 ly. Diameter = 160 ly.
Hind's Variable Nebula NGC 1555	04^h 22.9^m	+19°32^m			Neb	Taurus	Very faint reflection nebula.
Horsehead Nebula IC 434	05^h 41^m	−02°24'	11.3	60'	Neb	Orion	A dark nebula that resembles the shape of a horse head, positioned in front of a bright diffuse

Name	RA	Dec	Mag	Size	Type	Const	Comments
							nebula. This object is best seen using an H-Beta filter. Distance = 1500 ly.
Hubble Variable Nebula NGC 2261	06^h 39.2^m	+08°44'		2'	Neb	Monoceros	Gaseous nebula surrounding a variable star. Distance = 2600 ly. Diameter = 7 ly.
Hyades NGC 7845	04^h 27.0^m	+16°00'		4°	OC	Taurus	A very close open cluster with a triangular shape forming the head of the bull of the constellation Taurus. Distance = 150 ly. Diameter = 12 ly.
Lagoon Nebula M8	18^h 03.8^m	−24°23'	5.8	90'	Neb	Sagittarius	A very bright nebula which is an active star formation region. Distance = 5000 ly. Diameter = 100 ly.
Little Dumbbell NGC 650	01^h 42.3^m	+51°34'	12.0	4.8'	PN	Perseus	
NGC 891	02^h 22.6^m	+42°21'	10.0	13.5'	Gal	Andromeda	Edge-on spiral galaxy with prominent ban of dark interstellar matter along the galactic plane. Distance = 43 million ly. Diameter = 120 000 ly.
North American Nebula NGC 7000	20^h 58.8^m	+44°20'		120'	Neb	Cygnus	A diffuse nebula that resembles the outline of the North American continent. Best seen with an OIII filter. Distance 1600 ly. Diameter 45 ly.
Omega Cluster NGC 5139	13^h 26.8^m	−47°29'	3.7	36.3'	GC	Centaurus	The brightest and largest globular cluster in the sky. Best seen at lower or southern latitudes. Distance = 17 000 ly. Diameter = 100 ly.
Orion Nebula M42	05^h 35.4^m	−05°27'	4.0	66'	Neb	Orion	One of the most impressive diffuse nebulae in the sky. Visible to the naked eye bellow Orion's belt. Star forming region with four prominent stars called "The Trapezium" centered in the nebula. Distance = 1600 ly. Diameter = 30 ly.
Owl Nebula M97	11^h 14.8^m	+55°01'	11.2	3.2'	PN	Ursa Major	Also called M97. Irregular shaped planetary nebula. Distance = ~3000 ly. Diameter = 3 ly.
Pinwheel M33	01^h 33.9^m	+30°39'	5.7	62'	Gal	Triangulum	A large spiral galaxy structure with bright knots of dust and gas.

Name	RA	Dec	Mag	Size	Type	Const	Comments
							After the Andromeda Galaxy, it is the closest spiral galaxy to the Milky Way. Distance = 3 million ly. Diameter = 60 000 ly.
Pleiades M45	03^h 47.5^m	+24°07'	1.5	89'	OC	Taurus	Also called "The Seven Sisters" and M45. Bright, open cluster with 250 members. Six or seven stars easily visible to the naked eye. Distance = 450 ly. Diameter = 20 ly.
Ring Nebula M57	18^h 53.6^m	+33°02'	9.0	2.5'	PN	Lyra	A very famous planetary nebula with an extremely hot blue dwarf star in the center. Distance = 1400 ly. Diameter = 0.5 ly.
Ring-Tail Galaxy NGC 4038	12^h 01.9^m	−18°53'	13.0	3.2'	Gal	Corvus	A very unusual pair of interacting galaxies resulting in extraordinary long, curved filaments extending from each galaxy. Distance = 90 million ly. Diameter = 100 000 ly.
Rosette Nebula NGC 2244	06^h 30.3^m	+05°03'			Neb	Monoceros	A diffuse nebula surrounding an open cluster. Clouds of dark interstellar material against the glowing back ground of the nebula give it its "rose" appearance. Distance = 2600 ly. Diameter = 55 ly.
Saturn Nebula NGC 7009	21^h 04.2^m	−11°22'	8.0	1.7'	PN	Aquarius	Planetary nebula with an elongated ring resembling the rings of Saturn. Distance = 4000 ly. Diameter = 0.5 ly.
Sculptor Galaxy NGC 253	00^h 47.6^m	−25°17'	7.1	25.1'	Gal	Sculptor	The brightest member of the Sculptor Galaxy Cluster and very similar in size to our Milky Way. Total luminosity of 5 billion suns. Distance = 7.5 million ly. Diameter = 70 00 ly.
Siamese Twins NGC 4567	12^h 36.5^m	+11°15'	11.3	3.0'	Gal	Virgo	Elongated galaxy with close companion galaxy.
Sombrero Galaxy M104	12^h 40^m	−11°37'	8.3	8.9'	Gal	Virgo	Also called M104. Edge-on galaxy with a dark band of interstellar

Name	RA	Dec	Mag	Size	Type	Const	Comments
							matter in the equatorial galactic plane resembling the brim of a sombrero. Distance = 30 million ly. Diameter = 82 000 ly.
Spindle Galaxy NGC 3115	10^h 05.2^m	–07°43'	9.2	8.3'	Gal	Sextans	A greatly flattened, elliptical galaxy shaped like a spindle. Distance = 27 million ly. Diameter = 30 000ly
Stephan's Quintet NGC 7317	22^h 36.1^m	+33°57'	12.7	2.2'	Gal	Pegasus	A group of five galaxies, some of which are connected by "bridges" of interstellar matter.
Sunflower Galaxy M63	13^h 15.8^m	+42°02'	8.6	12.3'	Gal	Canes Venatici	An elongated galaxy with a bright core and a tight spiral pattern resembling a giant celestial flower. Distance = 35 million ly. Diameter = 90 000 ly.
Swan Nebula M17	18^h 20.8^m	–16°11'	6.0	46'	Neb	Sagittarius	Diffuse nebula resembling a swan swimming across water. Distance = 5500 ly. Diameter = 15 ly.
Tarantula Neb NGC 2070	05^h 38.6^m	–69°05'	8.2	40'	Neb	Dorado	A giant diffuse nebula located in the Large Magellanic Cloud is the largest object of its kind known in the Universe. Distance = 169 000ly. Diameter = 800ly
Trifid Nebula M20	18^h 02.3^m	–23°02'	6.3	29'	Neb	Sagittarius	A diffuse nebula characterized by three dark bands dividing the bright regions of the nebula. Distance = 3700 ly.
Veil Nebula NGC 6992 / 6960	20^h 56.4^m	+31°43'		2.5°	SR	Cygnus	A beautiful complex nebula, whose delicate filament structure is best seen using an OIII filter. The Veil Nebula is the remnant of an ancient supernova that exploded some 35 000 years ago. Distance = 1500 ly.
Whirlpool Galaxy M51	13^h 29.9^m	+47°12'	8.4	11'	Gal	Canes Venatici	Also called M51. One of the closest and brightest galaxies with very pronounced spiral arms. Distance = 35 million ly. Diameter = 100 000 ly.
Wild Duck Cluster M11	18^h 51.1^m	–06°16'	5.8	14'	OC	Scutum	Also called M11. A rich open cluster containing about 500 stars. Distance = 6000 ly. Diameter = 21 ly.

Asterisms

A collection of unique star patterns.

Name	RA	Dec	Const	Comments
Kite Cluster	01^h 40^m	+59°00'	Cas	A diamond-shaped pattern of stars resembling a kite with a string of 5 stars making up the tail.
Triangle Cluster	01^h 54^m	+38°00'	And	A long string of stars leading to an isosceles triangle situated near the bright, open cluster NGC 752.
Perseus Cluster	03^h 28^m	+49°00'	Per	Called the Perseus Moving Cluster, this asterism is a large association of over 50 stars scattered across 3° of the constellation Perseus. All stars visible are bright, very young, hot stars dating back only about 50 million years.
Kemble's Cascade	03^h 57^m	+63°00'	Cam	Long strand of 5th- to 8th-magnitude stars stretching over 2°. Continuing southeast from the cluster will lead to the open cluster NGC 1502.
Little Fish	05^h 18^m	+33°30'	Aur	More than a dozen stars make up this cluster which cover an area of 30' × 75' across
Arrowhead Cluster	06^h 40^m	−09°00'	CMa	A distinctive V-shaped pattern of star resembling an arrowhead positions about 7° north of the star Sirius.
Number 3 Cluster	06^h 53^m	−10°12'	CMa	A pattern of faint stars seen as a backwards number "3" through a low power eyepiece. Look for the bright, open cluster M50 just northwest of this asterism.
Crown Cluster	10^h 50^m	+56°12'	UMa	A small "crown" of stars with a bright "jewel" star near the center located 2° west of the star Merak.
Coma Star Cluster	12^h 22^m	+25°48'	CmB	The Coma Star Cluster. A V-shaped pattern of stars visible to the naked eye. This cluster is one of the closest to Earth at a distance of 250 ly.
Star Gate Cluster	12^h 36^m	−12°00'	Crv	A unique triangle of stars with another triangle of fainter stars in its center.
Mini-Coathanger	16^h 30^m	+80°12'	UMi	A pattern of 10 faint stars spanning $\frac{1}{2}$ degree, nicknamed the "Mini-coathanger" or "S" Cluster
S Cluster	16^h 37^m	+31°00'	Her	A curving pattern of a dozen stars resembling a backwards letter "S".
Ring Cluster	18^h 04^m	+26°30'	Her	A faint ring of stars with a brighter 7th-magnitude star inside the ring.
W Cluster	18^h 35^m	+72°18'	Dra	An asterism of 20 stars, with the 5 brightest stars making a distinctive "W" pattern $\frac{1}{2}$ degree across.

Name	RA	Dec	Const	Comments
Coathanger Cluster	19^h 25^m	+20°12'	Vul	Called "Brocchi's cluster" or the "Coathanger cluster". Visible to the naked eye 8° south of the star Albireo. Ten visible stars make up a coat hanger shape spanning 1° across.
Spiral Cluster	20^h 14^m	+36°30'	Cyg	A sprawling pattern of stars resembling the arms of a spiral galaxy with a brighter reddish star in the center.
Diamond Cluster	20^h 39^m	+13°30'	Del	Seen as a bright triangle of 7th-magnitude stars, with 6th-magnitude Theta Delphini being the brightest.
Horseshoe Cluster	21^h 07^m	+47°18'	Cyg	A "horseshoe" pattern of over a dozen 7th- to 11th-magnitude stars measuring 20' across lying near the bright star Deneb.
Number 7 Cluster	23^h 07^m	+59°00'	Cas	Nine 5th- and 6th-magnitude stars making up a sideways number "7".
Arrow Cluster	23^h 21^m	+62°30'	Cas	One degree of stars resembling an arrow pattern northeast of the cluster M52.

CCD Objects

Only available in the NexStar 8/11 GPS and NexStar 5i/8i. A collection of galaxy clusters well suited to CCD imaging.

Name	RA	Dec	Const	Comments
NGC 383 Group	01^h 07^m 23^s	+32°25'	Pisces	A well-populated galaxy group with more than a dozen small galaxies spread over a 1.5° area. A good object for f/2 Fastar imaging.
Arp 318	02^h 09^m 24^s	−10°08'	Cetus	A string of four spiral galaxies arranged in a line curving east and south.
NGC 1068 & 1055	02^h 42^m 24^s	+00°13'	Cetus	This galaxy pair consists of a circular spiral galaxy with a brilliant core and less bright spiral arms, with a visible dust lane running through one of its arms.
NGC 1097A	02^h 46^m 12^s	−30°14'	Fornax	A strongly barred spiral galaxy with a large bright oval core. Detached from the galaxy is a 30' companion off the northern spiral bar.
Fornax Cluster	03^h 22^m 42^s	−37°12'	Fornax	This cluster is centered around its brightest member, NGC 1316, a bright, slightly elongated oval galaxy surrounded by many fainter galaxies.

Name	RA	Dec	Const	Comments
NGC 1400 & 1407	03^h 39^m 59^s	−18°35'	Eridanus	Two very similar spiral galaxies with bright cores and faint halos separated by 12'.
NGC 1532 & 1531	04^h 12^m 06^s	−32°52'	Eridanus	This visually contrasting galaxy pair is separated by only 1.75'. NGC 1532 is a 10th-magnitude elongated galaxy with a bright core. NGC 1531 is a smaller, less elongated companion galaxy.
NGC 1723 Group	04^h 59^m 00^s	−11°00'	Eridanus	A group of four galaxies in the shape of a cat's paw with three faint galaxies in an arc and the brighter NGC 1723, directly north of the trio.
Bode's Galaxies	09^h 55^m 36^s	+69°20'		This galaxy pair made up of M81 and M82 are both bright, oval galaxies separated by half of a degree. Good object for Fastar f/2 imaging.
NGC 3158 Group	10^h 13^m 48^s	+38°46'		This galaxy group centered around NGC 3158 is made up of eight galaxies that span a 10' diameter and can be imaged at f/6.3 or at f/2 with a Fastar lens.
NGC 3190 Group	10^h 18^m 06^s	+21°50'	Leo	NGC 3190 is the brightest member of this trio of galaxies, which include NGC 3187 and NGC 3193.
Hydra I Cluster	10^h 36^m 42^s	−27°32'	Hydra	Hydra I is a cluster of eight galaxies that lie at a distance of 200 million light-years away. All are 11th- and 12th-magnitude galaxies, the farthest being about 30' from each other. Good object for f/2 Fastar imaging.
M105 Trio	10^h 47^m 48^s	+12°35'	Leo	This galaxy group is composed of an 8 arc minute triangle of contrasting galaxies including an elliptical, lenticular and loose spiral galaxy.
Butterfly Galaxy	10^h 50^m 00^s	+33°00'		Two elongated interacting galaxies connected at their tips giving the appearance of a butterfly.
Leo's Triple	11^h 19^m 30^s	+13°15'	Leo	The Leo Triple is comprised of two bright Messier objects (M65 and M66) and NGC 3628 that form a triangle of more than 30' across. Best suited for Fastar f/2 imaging.
Copeland's Septet	11^h 37^m 54^s	+21°59'	Leo	Copeland's Septet is a tight cluster of 6 faint galaxies scattered across a 6' field of view. The galaxies range in brightness from magnitude 13.6 to 15. This cluster is best imaged through an f/6.3 reducer.
Ring-Tail Galaxy	12^h 01^m 53^s	−18°53'	Corvus	NGC 4038-39 is an unusual pair of interacting galaxies that appear as one comma-shaped object.

Name	RA	Dec	Const	Comments
NGC 4298 & 4302	12ʰ 21ᵐ 43ˢ	+14°36'	Coma Berenices	This visually contrasting pair is only separated by 1' and is best imaged at f/6.3 with reducer lens.
Virgo Cluster	12ʰ 26ᵐ 12ˢ	+12°57'	Virgo	M86 lies at the center of this galaxy cluster surrounded by seven satellite galaxies of various sizes and magnitude.
Siamese Twins	12ʰ 36ᵐ 36ˢ	+11°14'	Virgo	The Siamese Twins are a pair of interacting galaxies that appear to be joined at the tips.
NGC 5350 Group	13ʰ 53ᵐ 24ˢ	+40°22'	Canes Venatici	This group is positioned near a 6.5-magnitude star and has a bright spiral galaxy, NGC 5471 less than 0.5° to the northeast. A good object for f/2 Fastar imaging.
NGC 5774-75	14ʰ 54ᵐ 00ˢ	+03°33'	Virgo	Two spiral galaxies, one seen edge-on, the other seen face-on, with a faint companion galaxy to the south.
Stephan's Quintet	22ʰ 36ᵐ 07ˢ	+33°57'	Pegasus	These five galaxies all lie within a 3.5' area and range in brightness from 12.6 to 13.6. Stephan's Quintet is positioned just south of another good CCD object NGC 7330.
Deer Lick Cluster	22ʰ 37ᵐ 06ˢ	+34°25'	Pegasus	NGC 7331 is a nearly edge-on spiral galaxy with four smaller companion galaxies orbiting towards the east.
Pegasus I Cluster	23ʰ 20ᵐ 42ˢ	+08°13'	Pegasus	This galaxy cluster is made up of more than 10 faint galaxies ranging in magnitude from 11 to 13. The cluster spans 30' across with two equal brightness twin galaxies in its center. Pegasus I is a good object for Fastar imaging.
NGC 7782 Group	23ʰ 53ᵐ 54ˢ	+07°58'	Pisces	At 12th-magnitude NGC 7782 is the brightest of this five galaxy cluster located in Pisces.

Stars

In addition to the Named, Double, and Variable star lists, additional stars are accessible using the **Star** button on the hand control. The number of stars varies by NexStar model. Lists of the stars are available in the Downloads section of my NexStar Resource Site – http://www.NexStarSite.com.

- **NexStar 60/80/114/4** – 2824 stars. The new GT hand control organizes this list by SAO catalog numbers. The original GT hand control uses a NexStar-specific number that can be cross-referenced to the SAO catalog with the list available on my web site.

- **NexStar 5/8** – 10 384 stars organized by a NexStar-specific number that can be cross-referenced to the SAO catalog with the list available on my web site.

- **NexStar 5i/8i and 8/11 GPS** – 29 522 stars organized by SAO catalog numbers.

Messier Objects

All 110 Messier objects are available in all NexStar models. The Messier objects are listed in Chapter 6 and a list is also available in the Downloads section of my NexStar Resource Site.

Caldwell Objects

All 109 Caldwell objects are available in all NexStar models. A list of the Caldwell objects is available in the Downloads section of my NexStar Resource Site.

NGC Objects

The NexStar 60/80/114/4 hand control includes 1165 NGC objects. All other models include the entire NGC catalog of 7840 objects. Both lists are available in the Downloads section of my NexStar Resource Site.

IC Objects

The NexStar 5i/8i and 8/11 GPS contain the entire IC catalog of 5386 objects. A list of the IC objects is available in the Downloads section of my NexStar Resource Site.

Abell Objects

The NexStar 5i/8i and 8/11 GPS contain the first 2712 objects in the Abell galaxy cluster catalog. The magnitude given by the hand control is the magnitude of the tenth-brightest galaxy in the cluster. A list of these Abell objects is available in the Downloads section of my NexStar Resource Site.

PC and Palmtop Software Compatible with NexStar Telescopes

There are a wide variety of software packages available for controlling your NexStar telescope. Programs written for Windows, Macintosh, and palmtop computers (PDAs) can provide extended control and enhanced features for your scope. Prices range from free to a few hundred dollars. While many of the packages are similar in function, some are quite unique. I recommend you spend some time researching at the manufacturers' web sites to determine which software package or packages best suit your needs.

Table C.1 shows the designations used to denote which models of NexStar scopes are compatible with each software package.

Following is a list of software packages available at the time of this writing; additional software is introduced continually. Please visit the PC Control section of the NexStar Resource Site – http://www.NexStarSite.com – for the most up-to-date list. All information is subject to change by the software manufacturer; please visit the web site listed for the latest details.

Table C.1. Designations denoting compatibility of models of NexStar scopes with software packages

Model	Description
Original GT	The original GT (computerized GoTo) hand control for the NexStar 60/80/114/4 telescopes, as well as all Tasco StarGuide telescopes.
New GT	The upgraded GT hand control for the NexStar 60/80/114/4 telescopes. This hand control is included with all little NexStars manufactured after December 2001. See the sidebar Old versus New GT Hand Control in Chapter 3 to determine whether you have the original or new GT hand control.
NexStar 5/8	The original NexStar 5 and 8 models. These models were discontinued in the summer of 2002 and replaced with the NexStar 5i and 8i.
NexStar 5i/8i	The new NexStar 5i and 8i telescopes. Only the optional computerized hand control allows PC control.
NexStar GPS	The NexStar 8/11 GPS telescopes.

NexStar Observer List

Author: Michael Swanson
Cost: Free
Web Site: http://www.NexStarSite.com/
Computer systems supported: Windows 95/98/ME/NT/2000/XP
Models of NexStar supported: Original GT, New GT, NexStar 5/8, NexStar 5i/8i, NexStar GPS

Arrow Keys for GuideStar

Authors: Michael Ganslmeier and Matthias Bopp
Cost: Free
Web Site: http://www.dd1us.de
Computer systems supported: Windows 2000/XP
Models of NexStar supported: HC models (the manual hand control little NexStars) and GT models with modifications
Note: A unique package that allows the PC to act as the direction buttons on the hand control. This allows complete remote control of the little NexStars after the initial alignment is accomplished. Visit the site and download the article on "Controlling the NexStar 4/60/80/114 series telescopes fully remotely" for complete details.

ASCOM

Publisher: Astronomy Common Object Model (ASCOM)
Cost: Free
Web Site: http://ascom-standards.org
Computer systems supported: Windows 95/98/ME/NT/2000/XP
Models of NexStar supported: Original GT, New GT, NexStar 5/8, NexStar 5i/8i, NexStar GPS
Note: This is a plug-in that is used by other programs to control a telescope, not a standalone program.

AstroPlanner

Publisher: iLanga, Inc.
Cost: $20 download version; $30 CD-ROM version
Web Site: http://www.ilangainc.com/astroplanner
Computer systems supported: Mac OS and Windows 95/98/ME/NT/2000/XP
Models of NexStar supported: Original GT, New GT, NexStar 5/8, NexStar 5i/8i, NexStar GPS

Cartes du Ciel (Sky Charts)

Author: Patrick Chevalley
Cost: Free
Web Site: http://www.stargazing.net/astropc
Computer systems supported: Windows 95/98/ME/NT/2000/XP
Models of NexStar supported: Uses the ASCOM plug-in above and supports the same scopes. Note that it is necessary to download the ASCOM platform from the ASCOM link above **and** you must download the ASCOM driver from the Cartes du Ciel web site – the "Complete Package" download does not include either.

Deepsky 2003

Author: Steven Tuma
Cost: $53 on CD, $40 when downloaded
Web Site: http://www.deepsky2000.net
Computer systems supported: Windows 95/98/ME/NT/2000/XP
Models of NexStar supported: Uses the ASCOM plug-in above and supports the same scopes.

Pocket Deepsky

Author: Steven Tuma
Cost: $20 when purchased separately, free with a purchase of Deepsky 2003
Web Site: http://www.deepsky2000.net
Computer systems supported: Windows CE Version 3
Models of NexStar supported: NexStar 5/8

Desktop Universe

Publisher: Main Sequence Software
Cost: $300
Web Site: http://www.desktopuniverse.com
Computer systems supported: Windows 95/98/ME/NT/2000/XP
Models of NexStar supported: Uses the ASCOM plug-in above and supports the same scopes.

DigitalSky Voice

Publisher: Astro Physics, Inc.
Cost: $175
Web Site: http://www.digitalskyvoice.com
Computer systems supported: Windows 95/98/ME/NT/2000/XP
Models of NexStar supported: New GT, NexStar 5/8, NexStar 5i/8i, NexStar GPS

Earth Centered Universe Pro

Publisher: Nova Astronomics
Cost: $60
Web Site: http://www.nova-astro.com
Computer systems supported: Windows 3.1/95/98/ME/NT/2000/XP
Models of NexStar supported: Uses the ASCOM plug-in above and supports the same scopes.

Guide CD-ROM Star Chart, Version 8

Publisher: Project Pluto
Cost: $89
Web Site: http://www.projectpluto.com
Computer systems supported: DOS, Windows 3.1/95/98/ME/NT/2000/XP
Models of NexStar supported: Original GT, New GT, NexStar 5/8, NexStar 5i/8i, NexStar GPS

HNSKY

Author: Han Kleijn
Cost: Free
Web Site: http://www.hnsky.org
Computer systems supported: Windows 95/98/ME/NT/2000/XP
Models of NexStar supported: Uses the ASCOM plug-in above and supports the same scopes. In addition to the ASCOM plug-in, you must also download the HNSKY to ASCOM Interface from the HNSKY web site.

MegaStar 5

Publisher: Willmann-Bell, Inc.
Cost: $130
Web Site: http://www.willbell.com/software/megastar/index.htm
Computer systems supported: Windows 95/98/ME/NT/2000/XP
Models of NexStar supported: New GT, NexStar 5/8, NexStar 5i/8i, NexStar GPS

NGCView

Publisher: Rainman Software
Cost: $90
Web Site: http://www.rainman-soft.com
Computer systems supported: Windows 95/98/ME/NT/2000/XP
Models of NexStar supported: Uses the ASCOM plug-in above and supports the same scopes.

Observer

Publisher: Procyon Systems
Cost: $90 for Intermediate, $125 for Advanced, Basic does not support telescope control
Web Site: http://www.procyon-sys.com
Computer systems supported: Power Mac
Models of NexStar supported: New GT, NexStar 5i/8i, NexStar GPS

Planetarium

Author: Andreas Hofer
Cost: $24
Web Site: http://www.aho.ch/pilotplanets
Computer systems supported: Palm Pilot (or compatible) running Palm OS version 2.0 or higher
Models of NexStar supported: Original GT, New GT, NexStar 5/8, NexStar 5i/8i, NexStar GPS – requires download of telescope control plug-in for specific scopes

Real-Time PC Software for Astronomical Observers

Author: Robert Sheaffer
Cost: Free
Web Site: http://www.debunker.com/astro/rtastro.html
Computer systems supported: DOS, Windows 3.1/95/98/ME
Models of NexStar supported: Original GT, New GT, NexStar 5/8, NexStar 5i/8i, NexStar GPS

Satellite Tracker

Author: Brent Boshart
Cost: $20
Web Site: http://www.heavenscape.com
Computer systems supported: Windows 95/98/ME/NT/2000/XP
Models of NexStar supported: New GT, NexStar 5/8, NexStar 5i/8i, NexStar GPS – continuous tracking on the NexStar GPS (HC 1.6 or higher) and NexStar 5i/8i; leapfrog tracking on other models

SkyChart III

Publisher: Southern Stars Systems
Cost: $50
Web Site: http://southernstars.com/skychart
Computer systems supported: Windows 95/98/ME/NT/2000/XP and Mac
Models of NexStar supported: NexStar 5/8 and reportedly supports New GT, NexStar 5i/8i, and NexStar GPS when the Ultima 2000 driver is selected

SkyMap Pro

Publisher: SkyMap Software
Cost: $100
Web Site: http://www.skymap.com
Computer systems supported: Windows 95/98/ME/NT/2000/XP
Models of NexStar supported:
Version 7 supports the NexStar 5/8; version 8 supports New GT, NexStar 5/8, NexStar 5i/8i, NexStar GPS

StarParty

Publisher: PFD Systems
Cost: $12
Web Site: http://www.pfdsystems.com
Computer systems supported: Palm Pilot or compatible – see web site for OS version requirements
Models of NexStar supported: Original GT, New GT, NexStar 5/8, NexStar 5i/8i, NexStar GPS

STAR Atlas:PRO

Publisher: SKYLab Astronomy Software
Cost: $130
Web Site: http://www.skylab.com.au
Computer systems supported: Windows 95/98/ME/NT/2000/XP
Models of NexStar supported: Uses the ASCOM plug-in above and supports the same scopes

Starry Night Pro

Publisher: Space Software
Cost: $180
Web Site: http://www.starrynight.com
Computer systems supported: Current version (version 4) only supports Windows 98/ME/2000/XP and Mac OS X 10.1 or higher. Version 3 supports Windows 95/98/ME/NT/2000/XP and older Mac OS versions.
Models of NexStar supported:
Windows version – can use the ASCOM plug-in for controlling models listed in the ASCOM section above. Mac version – Various plug-ins are available supporting New GT, NexStar 5/8, NexStar 5i/8i, NexStar GPS

TheSky Level II or Higher

Publisher: Software Bisque
Cost: Level II – $130
Web Site: http://www.bisque.com
Computer systems supported: Windows 95/98/ME/NT/2000/XP and Mac
Models of NexStar supported: New GT, NexStar 5/8, NexStar 5i/8i, NexStar GPS

TheSky Pocket Edition

Publisher: Software Bisque
Cost: $50
Web Site: http://www.bisque.com
Computer systems supported: Windows CE Version 3 on Pocket PC, Windows CE 2.11 for Palm-size PC and a few other version/hardware combinations detailed on the web site.
Models of NexStar supported: New GT, NexStar 5/8, NexStar 5i/8i, NexStar GPS

Voyager III with SkyPilot

Publisher: Carina Software
Cost: $200 (be certain to purchase SkyPilot, sold separately)
Web Site: http://www.carinasoft.com
Computer systems supported: Windows 98/ME/2000/XP
Models of NexStar supported: NexStar 5/8, NexStar GPS, New GT, NexStar 5i/8i

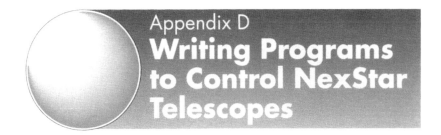

Appendix D
Writing Programs to Control NexStar Telescopes

In addition to the information presented here, several resources are available at the NexStar Resource Site (http://www.NexStarSite.com) to include sample programs written in Visual Basic 6, a widely used programming platform. The complete source code for these programs is available for download from the NexStar Resource Site. The source code requires that you have Visual Basic installed on your computer.

Basic Control Commands – Get and GoTo

For many programs, these are the only commands that will be necessary. The GoTo commands direct the telescope to slew to the coordinates provided. The Get commands query the telescope for the current coordinates of the position the telescope is pointed. There are commands for both RA–Dec and Alt–az coordinates.

The communications parameter for all models is 9600 bits per second, no parity, no stop bit, and no flow control. This last parameter causes many problems, be sure you **do not** select **Xon/Xoff** (sometimes called software flow control) or **Hardware** flow control.

As described in Chapter 10, various models of NexStar telescopes have been introduced with differing command sets. The following sections discuss the commands for each of the three distinct protocols existing in the NexStar line.

NexStar 5/8

GoTo RA–Dec

- PC sends binary value 63 (ASCII "?").
- The telescope replies with binary value 35 (ASCII "#").
- PC sends binary value 82 (ASCII "R") followed by the RA high-byte, RA low-byte, Dec high-byte, Dec low-byte.
- The telescope slews to the RA–Dec coordinates – after completion the telescope replies with binary value 64 (ASCII "@").

Get RA–Dec

- PC sends binary value 63 (ASCII "?").
- The telescope replies with binary value 35 (ASCII "#").
- PC sends binary value 69 (ASCII "E").
- The telescope replies with the RA high-byte, RA low-byte, Dec high-byte, Dec low-byte.

GoTo Alt–az

- PC sends binary value 63 (ASCII "?").
- The telescope replies with binary value 35 (ASCII "#").
- PC sends binary value 65 (ASCII "A") followed by the Azm high-byte, Azm low-byte, Alt high-byte, Alt low-byte.
- The telescope slews to the Alt–az coordinates – after completion the telescope replies with binary value 64 (ASCII "@").

Get Alt–az

- PC sends binary value 63 (ASCII "?").
- The telescope replies with binary value 35 (ASCII "#").
- PC sends binary value 90 (ASCII "Z").
- The telescope replies with the Azm high-byte, Azm low-byte, Alt high-byte, Alt low-byte.

Note that the high-byte and low-byte values for these commands are binary values formatted as individual ASCII characters. For example, to send a GoTo command with RA high-byte 124 and low-byte 54 and Dec high-byte 15 and low-byte 32, the correct Visual Basic string would be:

```
"R" & Chr(124) & Chr(54) & Chr(15) & Chr(32)
```

To calculate the high-byte, divide the encoder value by 256 and convert to an integer (drop the fractional part). To calculate the low-byte, divide the encoder value by 256 and use the remainder (modulus operation or the fractional part multiplied by 256). For programming examples and other details, refer to the NexStar Resource Site.

Original GT

GoTo RA–Dec

- PC sends binary value 82 (ASCII "R") followed by the RA high-byte, RA low-byte, binary value 0, Dec high-byte, Dec low-byte, binary value 0.
- The telescope slews to the RA–Dec coordinates.

Get RA–Dec

- PC sends binary value 69 (ASCII "E").
- The telescope replies with the RA high-byte, RA low-byte, binary value 0, Dec high-byte, Dec low-byte, binary value 0.

GoTo Alt–az

- PC sends binary value 65 (ASCII "A") followed by the Azm high-byte, Azm low-byte, binary value 0, Alt high-byte, Alt low-byte, binary value 0.
- The telescope slews to the Alt–az coordinates.

Get Alt–az

- PC sends binary value 90 (ASCII "Z").
- The telescope replies with the Azm high-byte, Azm low-byte, binary value 0, Alt high-byte, Alt low-byte, binary value 0.

Note that the high-byte and low-byte values for these commands are binary values formatted as individual ASCII characters. For example, to send a GoTo command with RA high-byte 124 and low-byte 54 and Dec high-byte 15 and low-byte 32, the correct Visual Basic string would be:

```
"R" & Chr(124) & Chr(54) & Chr(0) & Chr(15) & Chr(32)
                & Chr(0)
```

To calculate the high-byte, divide the encoder value by 256 and convert to an integer (drop the fractional part). To calculate the low-byte, divide the encoder value by 256 and use the remainder (modulus operation or the fractional part multiplied by 256). For programming examples and other details, refer to the NexStar Resource Site.

NexStar 8/11 GPS, NexStar 5i/8i, New GT

GoTo RA–Dec

- PC sends binary value 82 (ASCII "R") followed by the RA high-byte, RA low-byte, binary value 44 (ASCII ","), Dec high-byte, Dec low-byte.
- The telescope slews to the RA–Dec coordinates.

Get RA–Dec

- PC sends binary value 69 (ASCII "E").
- The telescope replies with the RA high-byte, RA low-byte, binary value 44 (ASCII ","), Dec high-byte, Dec low-byte.

GoTo Alt–az

- PC sends binary value 66 (ASCII "B") followed by the Azm high-byte, Azm low-byte, binary value 44 (ASCII ","), Alt high-byte, Alt low-byte.
- The telescope slews to the Alt–az coordinates.

Get Alt–az

- PC sends binary value 90 (ASCII "Z").
- The telescope replies with the Azm high-byte, Azm low-byte, binary value 44 (ASCII ","), Alt high-byte, Alt low-byte.

Note that the high-byte and low-byte values for these commands are binary values formatted as two-digit hexadecimal numbers sent as ASCII characters. For example, to send a GoTo command with RA high-byte 124 and low-byte 54 and Dec high-byte 15 and low-byte 32, the correct Visual Basic string would be:

```
"R" & Mid("00", 1, 2 - Len(Hex(124))) & Hex(124) & Mid("00", 1,
        2 - Len(Hex(54))) & Hex(54) & "," & Mid("00", 1,
    2 - Len(Hex(15))) & Hex(15) & Mid("00", 1, 2 - Len(Hex(32))) &
                        Hex(32)
```

To calculate the high-byte, divide the encoder value by 256 and convert to an integer (drop the fractional part). To calculate the low-byte, divide the encoder value by 256 and use the remainder (modulus operation or the fractional part multiplied by 256). For programming examples and other details, refer to the NexStar Resource Site.

Advanced Commands

Additional commands have been introduced for some of the NexStar models. The models applicable are noted for each of the commands below. Note that after many of these commands the telescope sends one extra byte – binary value 35 (ASCII "#") – to indicate the command has completed.

Get Hand Control Version

(New GT, NexStar 8/11 GPS, NexStar 5i/8i)

Returns hand control version only. Two bytes returned. Each byte is a single-byte binary number that should be converted to decimal. For example, in Visual Basic, use the ASC function to convert the binary value to decimal. The first byte is the whole-number version; the second byte is the fractional part. For example, if the values returned are 1 and 6, the hand control is version 1.6.

- PC sends binary value 86 (ASCII "V").
- The telescope replies with the whole-number-version byte and the fractional-number-version byte.

Cancel GoTo

(NexStar 8/11 GPS, NexStar 5i/8i)

Used to cancel a GoTo slew mid-progress.

- PC sends binary value 77 (ASCII "M").

Is GoTo in Progress

(NexStar 8/11 GPS, NexStar 5i/8i)

Used to determine if a GoTo is currently in progress. Returns a single-byte binary number that should be converted to decimal. For example, in Visual Basic, use the ASC function to convert the binary value to decimal.

- PC sends binary value 76 (ASCII "L").
- The telescope replies with a single-byte binary value. If the value is 49 (ASCII "1"), a GoTo is in progress. If the value is 48 (ASCII "0"), no GoTo is in progress.

Is Alignment Complete

(NexStar 8/11 GPS, NexStar 5i/8i)

Used to determine if the star alignment has been completed. The GoTo/Get RA–Dec commands do not function prior to alignment. Returns a single-byte binary number that should be converted to decimal. For example, in Visual Basic, use the ASC function to convert the binary value to decimal.

- PC sends binary value 74 (ASCII "J").
- The telescope replies with a single-byte binary value. If the value is 1, alignment is complete. If the value is 0, alignment has not been completed.

Tracking Mode (Start/Stop Tracking)

(NexStar 8/11 GPS hand control version 1.6 or higher, NexStar 5i/8i)

The Tracking Mode command provides the same capability as the same function found on the Menu button of the hand control. It is required via computer control when using the next command, Tracking Rate. Prior to sending a tracking rate change, you must use Tracking Mode to stop (turn off) tracking. To resume normal sidereal tracking you must use the Tracking Mode command to resume tracking. To resume tracking, you must choice one of three tracking modes: Alt–az, EQ North, or EQ South. Alt–az is the tracking mode used when the scope is mounted directly on the tripod. EQ North/South are used when the telescope is polar-aligned on a wedge (Northern or Southern Hemisphere). Unfortunately, there is no command to determine the current tracking mode, so a program that will use the Tracking Mode command must ask the user to select the correct mode.

- PC sends binary value 84 (ASCII "T") and a single-byte binary value to set the tracking mode as follows:
  0 – stop tracking
  1 – start Alt–az tracking
  2 – start EQ North tracking
  3 – start EQ South tracking.

Tracking Rate

(NexStar 8/11 GPS hand control version 1.6 or higher, NexStar 5i/8i)

The Tracking Rate command directs the telescope to begin slewing left, right, up, or down – the same directions available to the arrow buttons on the hand control. Providing you first stop tracking using the Tracking Mode command, the telescope will continue moving at the requested rate until you issue either another Tracking Rate command or another Tracking Mode command. For example, you can issue a Tracking Mode command to stop tracking, then a Tracking Rate command to slew up at 1° per second, and then after any amount of time, you can issue a Tracking Mode command to start Alt–az tracking and the telescope will resume normal tracking at sidereal rate. This method can be used to write programs that provide on-screen buttons to emulate the arrow buttons on the hand control.

Also note that you can start one axis tracking at any given rate and then start the other axis tracking at an independent rate. This allows tracking of objects, like satellites, moving on a predetermined course across the sky.

Following are the four variations of the Tracking Rate command to allow movement in any of the four directions. Each command requires 8 bytes; each byte shown is a decimal binary value.

Positive Azimuth Tracking (move right)
    80, 3, 16, 6, TrackRateHigh, TrackRateLow, 0, 0

Negative Azimuth Tracking (move left)
    80, 3, 16, 7, TrackRateHigh, TrackRateLow, 0, 0

Positive Altitude Tracking (move up)
    80, 3, 17, 6, TrackRateHigh, TrackRateLow, 0, 0

Negative Altitude Tracking (move down)
    80, 3, 17, 7, TrackRateHigh, TrackRateLow, 0, 0

Multiply your desired tracking rate, expressed in arc seconds per second, by a factor of 4. For example, 1° per second is 3600 arc seconds per second. Multiplied by four, this results in a tracking rate of 14 400. To calculate TrackRateHigh, divide this value by 256 and convert to an integer (drop the fractional part). To calculate TrackRateLow, divide the encoder value by 256 and use the remainder (modulus operation or the fractional part multiplied by 256). A sample program using these commands to emulate the hand control buttons is available for download from the NexStar Resource Site.

32-bit RA–Dec Get and GoTo

(NexStar 8/11 GPS hand control version 1.6 or higher, NexStar 5i/8i)

GoTo RA–Dec

- PC sends binary value 114 (ASCII "r") followed by the RA high-byte, RA mid-byte, RA low-byte, binary value 0, binary value 44 (ASCII ","), Dec high-byte, Dec mid-byte, Dec low-byte, binary value 0.
- The telescope slews to the RA–Dec coordinates.

Get RA–Dec

- PC sends binary value 101 (ASCII "e").
- The telescope replies with the RA high-byte, RA mid-byte, RA low-byte, binary value 0, binary value 44 (ASCII ","), Dec high-byte, Dec mid-byte, Dec low-byte, binary value 0.

Note that the high-byte, mid-byte, and low-byte values for these commands are binary values formatted as two-digit hexadecimal numbers sent as ASCII characters. For example, to send a GoTo command with RA high-byte 124, mid-byte 30, low-byte 54 and Dec high-byte 15, mid-byte 102, low-byte 32, the correct Visual Basic string would be:

```
"r" & Mid("00", 1, 2 - Len(Hex(124))) & Hex(124) & Mid("00", 1,
  2 - Len(Hex(30))) & Hex(30) & Mid("00", 1, 2 - Len(Hex(54))) &
    Hex(54) & "00," & Mid("00", 1, 2 - Len(Hex(15))) & Hex(15) &
      Mid("00", 1, 2 - Len(Hex(102))) & Hex(102) & Mid("00", 1,
                   2 - Len(Hex(32))) & Hex(32) & "00"
```

To calculate the high-byte, divide the encoder value by 65 536 and convert to an integer (drop the fractional part). To calculate the mid-byte, multiply the high-byte by 65 536 and subtract it from the encoder value, then divide the answer by 256 and convert to an integer. To calculate the low-byte, multiply the high-byte by 65 536, multiply the mid-byte by 256, and subtract both from the encoder value; the answer is the low-byte. For programming examples and other details, refer to the NexStar Resource Site.

Additional Commands

André Paquette has investigated the internal communications protocol of the NexStar 8/11 GPS and NexStar 5i/8i telescopes. Some of the more useful commands control the motor control and GPS modules. He has published an excellent guide to these commands for the technically inclined, which you will find available at his web site http://www.paquettefamily.ca/nexstar/

Appropriate care must be used with any of these additional commands since a few of them can render your telescope inoperable, requiring a return trip to Celestron (most likely at your expense!) to remedy the situation.

Appendix E

Glossary

AFOV See **apparent field of view.**

Airy disk The image of a star as seen through a telescope. Although a star is a pinpoint of light, due to the interaction of light and optics, it appears as a very small disk surrounded by one or more faint diffraction rings. Viewing the Airy disk requires good optics, steady seeing, and a well-collimated scope.

alt–azimuth or **altitude–azimuth** A method of mounting a telescope so that motion is allowed left to right (azimuth) and up and down (altitude).

apparent field of view (AFOV) The apparent angular slice shown by an eyepiece. A larger AFOV yields a "wider" view, while a narrow AFOV is often likened to viewing through a drinking straw. See also **field of view** and **true field of view.**

aperture The diameter of a telescope's objective (main lens or mirror). Larger apertures provide greater light-gathering power (the ability to show fainter objects) and higher resolution (the ability to show finer detail).

arc minute A unit (symbol ′) used to measure the angular separation of objects in the sky. An arc minute is $\frac{1}{60}$ of a **degree.**

arc second A unit (symbol ′′) used to measure the angular separation of objects in the sky. An arc second is $\frac{1}{60}$ of an **arc minute.**

asterism A recognizable pattern of stars, for example the Big Dipper or the Southern Cross.

asteroid A small rocky body orbiting the Sun or other star. In our solar system, great concentrations of asteroids are found in orbit between Mars and Jupiter.

averted vision A technique used to see faint objects. By looking to the side of the object, the faint light falls on the more sensitive part of the eye.

binary star See **double star.**

binocular viewers Often simply called "bino viewers", used with a telescope, this special adapter allows viewing with both eyes.

Caldwell catalog 109 **deep sky objects** to expand beyond the Messier list. Many of these objects are only visible to observers in the Southern Hemisphere. See also **Messier catalog.**

catadioptric telescope A telescope design using both mirror and lenses, for example, **Schmidt–Cassegrain** or **Maksutov–Cassegrain** telescopes.

celestial equator An imaginary line on the sky found directly above the Earth's equator. Objects north of the celestial equator have a positive **declination**; objects south have a negative **declination.**

celestial pole Either of the points in the sky directly above the Earth's North or South Pole.

collimation Alignment of the optical elements of a telescope, binocular, or other optical device.

comet A celestial body, made mostly of ice and rock, usually orbiting a star in a highly elliptical orbit.

compound telescope Another name, and easier to pronounce than **catadioptric**, for a telescope design using both mirror and lenses. **Schmidt-Cassegrain** or **Maksutov-Cassegrain** telescopes are examples.

corrector plate The large lens on the front of **Schmidt-Cassegrain** or **Maksutov-Cassegrain** telescopes. The corrector on a **Maksutov-Cassegrain** is also called the meniscus lens.

crescent Moon A lunar phase area when less than a quarter of the surface of the Moon visible from the Earth is illuminated.

DEC See **declination.**

declination A measurement, in degrees, of an object's angular distance from the **celestial equator**. Objects north of the **celestial equator** have a positive declination; objects south of the celestial equator have a negative declination. Used together with **right ascension**, provides the precise location of an object in the sky.

deep sky object Objects outside of our solar system are known collectively as deep sky objects or DSOs. DSOs range from individual stars to cities of stars known as galaxies. Refer to Chapter 2 for more details.

degree A unit (symbol °) used to measure the angular separation of objects in the sky. There are 360° measured around a complete circle.

double star A star that can be resolved with optical aid to show two stars. Multiple stars are also found with three or more members. Some are binary/multiple systems with the member stars revolving around a common center of gravity. Others are merely an optical effect with member stars separated by great distances but coincidentally in a common line of sight as viewed from the Earth.

DSO See **deep sky object.**

ecliptic The line in the sky that the planets, the Moon, and the Sun travel along. The line resembles a sine wave 23° above the **celestial equator** at its high point, 23° below the **celestial equator** at its low point. The course and seasonal orientation of the line is due to the 23° tilt of the Earth's axis and the Earth's path around the Sun.

ep or e.p. Sometimes used as an abbreviation for eyepiece.

equatorial mount A method of mounting a telescope so that one of the axes (called the **right ascension** axis) is aligned with the Earth's axis. The other axis (called the **declination** axis) is mounted at a 90° angle to the first. This allows easy tracking of the sky as the Earth rotates on its axis. The most common are fork-mounted scopes on a wedge or the German Equatorial Mount (GEM)

exit pupil The size, in millimeters, of the column of light projecting from the eyepiece of a telescope or binoculars. Refer to Chapter 2 for additional details.

field of view (FOV) The view in the eyepiece of a telescope or binoculars. Typical usage: "The Orion Nebula filled the entire field of view." Refer to Chapters 2 and 8 for more details.

focal length Traditionally, the measured distance from the main objective (lens or mirror) of an optical instrument, to the point where the image is brought to a focus. For compound telescopes such as **Schmidt-Cassegrain** and **Maksutov-Cassegrain**, the complex curves of the corrector and mirrors fold and compress the light path so that a relatively short optical tube projects a longer effective focal length.

focal ratio The result of dividing the **focal length** of a telescope by its **aperture**. Focal ratio is written "f/number". For example, the NexStar 80 has a focal length of 400 mm and an aperture of 80 mm resulting in a focal ratio of f/5 (400/80 = 5).

focal reducer An optical adapter that effectively decreases the **focal length** of a telescope. Sometimes called a telecompressor or reducer/corrector.

FOV See **field of view.**

full Moon The phase of the Moon when the entire illuminated portion of the Moon is facing the Earth.

galaxy A grouping of hundreds of millions or even billions of stars bound together by gravity and traveling together through space. There are billions of galaxies in the universe.

GEM See **equatorial mount.**

gibbous Moon A lunar phase area when more than half of the surface of the Moon visible from the Earth is illuminated.

globular cluster A group of hundreds of thousands of stars bound together by gravity.

IC catalog 5386 objects cataloged by J. L. E. Dreyer. In addition to individual and **double stars,** the majority of DSOs of interest to amateur astronomers are found in this or the **NGC catalog.**

local sidereal time The hour, minute, and second of **right ascension** directly overhead at any instant.

magnitude A measurement of the brightness of a celestial object as viewed from any given location (such as the Earth). Smaller numbers indicate brighter objects, with negative numbers being the brightest. Refer to Chapter 2 for more details.

Maksutov–Cassegrain telescope Sometimes called Maksutov or simply Mak, a telescope design using mirrors and a deeply curved **corrector plate** to create a compact instrument with a relatively long **focal length.** For a more detailed description and figure, refer to Chapter 2.

meniscus lens Also called corrector plate, the large lens on the front of **Maksutov–Cassegrain** telescopes.

meridian An imaginary line drawn from due south, directly overhead, and then due north.

Messier catalog 110 DSOs of various types of objects. Most of the brighter **deep sky objects** visible in the Northern Hemisphere are listed in Charles Messier's catalog. Generally these objects are the first DSO targets for beginning amateur astronomers. Messier 1 would be commonly referred to as "M1".

meteor A small particle that enters the Earth's atmosphere, burning away with a streak of light.

meteorite A small metal or rocky object that enters the Earth's atmosphere and reaches the surface before being completely consumed by the heat.

nebula A huge cloud of dust and/or gas found in space. Refer to Chapter 2 for more details.

new Moon The phase of the Moon when the entire illuminated portion of the Moon is facing away from the Earth. Thus, the unlit side of the Moon faces us.

Newtonian reflector telescope A telescope design using a large concave mirror and a small flat mirror to gather and focus light. For a more detailed description and figure, refer to Chapter 2.

NGC catalog 7840 objects cataloged by J. L. E. Dreyer. In addition to individual and **double stars,** the majority of DSOs of interest to amateur astronomers are found in this or the **IC catalog.**

open cluster A group of less than a few hundred stars loosely bound by their mutual gravity. Most are slowly drifting apart and are relatively young.

parfocal A characteristic of two or more eyepieces that allows them to be exchanged in a telescope with little or no refocusing required.

R/C reducer/corrector. See **focal reducer.**

RA See **right ascension.**

reducer/corrector See **focal reducer.**

reflector telescope A telescope design that utilizes a mirror for the main optical component. See **Newtonian reflector, Schmidt–Cassegrain,** and **Maksutov–Cassegrain** for examples.

refractor telescope The oldest telescope design, a refractor uses a lens for the main optical component. For a more detailed description and figure, refer to Chapter 2.

resolution Smallest detail able to be detected in an optical device. Usually measured in **arc seconds.**

right ascension Lines drawn against the sky running from the north **celestial pole** to the south **celestial pole.** Abbreviated RA. Right ascension changes as we move "east" and "west" in the sky. Right ascension is measured in hours, minutes, and seconds. The entire sky is divided into 24 hours, each hour is divided into 60 minutes, and each minute is divided into 60 seconds. Used together with **declination**, provides the precise location of an object in the sky.

RS-232 A communications standard in the computer industry. The port on the bottom of the NexStar hand control is RS-232-compatible. The computer industry is replacing RS-232 with newer standards such as **USB**. See Chapter 10 for more details.

Schmidt–Cassegrain telescope (SCT) A telescope design using mirrors and a relatively flat **corrector plate** to create a compact instrument with a long **focal length**. For a more detailed description and figure, refer to Chapter 2.

SCT See **Schmidt–Cassegrain telescope.**

seeing conditions The overall quality of the local atmosphere as it affects celestial observations. The three main factors affecting seeing conditions are known as transparency, seeing, and light pollution. Transparency is affected by particles in the air such as moisture, dust, pollution, etc. Seeing is the effect of unsteady air – basically air turbulence. Light pollution is caused by the enormous amount of light civilization sends into the night sky. The brightness of the Moon has a similar affect. For more details, refer to Chapter 2.

serial port A name commonly used for an **RS-232** port on a computer. See Chapter 10 for more details.

sidereal rate The rate at which the stars move across the sky. Each star takes a little less than 24 hours to make a complete circuit around the sky.

slew Technically, any movement of a telescope upon its mount, but more commonly used to indicate moving a scope with the motors on the mount.

terminator The line dividing the lighted and dark halves of a planetary body. Generally considered most important with the Moon since lunar features are most striking along the terminator.

TFOV See **true field of view.**

transparency See **seeing conditions.**

true field of view (TFOV) The actual angular distance across the **field of view** in an eyepiece. In any given telescope, the true field of view changes as you use various eyepieces. Refer to Chapters 2 and 8 for more details.

USB – Universal Serial Bus The modern replacement for the RS-232 serial port. All new computers are equipped with USB ports and many no longer come with an RS-232 port. Since the interface on NexStar telescopes is RS-232, some newer computers will require a "USB to serial" adapter. See Chapter 10 for more details.

variable star A star that changes magnitude (brightness) by a measurable degree. Some are caused by unstable nuclear reactions. Others are actually binary or **double star** systems resulting in lesser magnitude with both stars aligned with our line of sight.

vignette Darkening around the edges of the **field of view**, usually most noticeable in astrophotographs. Generally this is caused by using an equipment combination that exceeds the maximum possible field of view for a telescope.

visual back The 1.25-inch or 2-inch adapter on the back of a **Schmidt–Cassegrain** or **Maksutov–Cassegrain** telescope. Typically a diagonal is placed into the visual back to allow comfortable viewing.

waning Moon Any phase of the Moon after the **full Moon** and prior to the **new Moon**. Thus the Moon is "shrinking" in phase each night.

waxing Moon Any phase of the Moon after the **new Moon** and prior to the **full Moon**. Thus the Moon is "growing" in phase each night.

wedge An adapter placed between the tripod and a fork-mounted telescope to allow polar alignment.

zenith The point directly overhead.

Index